Refugees, Recent Migrants and Employment

Routledge Research in Population & Migration

SERIES EDITOR: PAUL BOYLE, *University of St. Andrews*

Refugees, Recent Migrants and Employment

Challenging Barriers and Exploring Pathways

Edited by Sonia McKay

Routledge
Taylor & Francis Group
New York   London

First published 2009
by Routledge
711 Third Avenue, New York, NY 10017

Simultaneously published in the UK
by Routledge
2 Park Square, Milton Park, Abingdon, Oxfordshire OX14 4RN

Routledge is an imprint of the Taylor & Francis Group, an informa business

First issued in paperback 2011

© 2009 Taylor & Francis

Typeset in Sabon by IBT Global.

Library of Congress Cataloging in Publication Data
Refugees, recent migrants and employment : challenging barriers and exploring pathways / edited by Sonia McKay.
p. cm. -- (Routledge research in population & migration ; 11)
Includes bibliographical references and index.
ISBN 978-0-415-98877-3
1. Alien labor.   2. Refugees--Employment.   3. Immigrants--Employment.
I.   McKay, Sonia.

HD6300.R44 2009
331.6'2--dc22   2008012821

ISBN13: 978-0-415-98877-3 (hbk)
ISBN13: 978-0-415-80786-9 (pbk)
ISBN13: 978-0-203-89074-5 (ebk)

Contents

Acknowledgments

This book would not have been possible without the involvement of the many hundreds of migrants and refugees who have generously given their time to the book's contributors, by participating in in-depth interviews or in completing questionnaires. The stories of their struggles in their new destinations and of their determination to overcome the hurdles that are placed in their way, while inspiring, nevertheless starkly expose the real consequences of globalism.

I would like to thank all of the book's contributors who retained their enthusiasm for this project through the many months of its gestation.

Finally I would also like to thank all of my colleagues at the Working Lives Research Institute for their support and encouragement, but in particular, those researchers who have worked with me on the migrant and refugee projects: Necla Acik, Deepta Chopra, Marc Craw, Sukhwant Dhaliwal, Amar Dhudwar, Siddig Elzailaee, Eugenia Markova, Anna Paraskevopoulou, Andrea Winkelmann-Gleed and Tessa Wright.

Introduction

The aim of this book has been to draw together a wide range of contributions reflecting the experiences and impact of migration in a number advanced capitalist economies in an era of globalisation and in a period when, almost universally, state policies have focused on restricting migration and limiting the rights of entry of refugees. The contributors examine migration from the perspectives of migrants and refugees themselves. It is their needs, and consequently the obligations of host countries to respond to them, that the contributors address. Migration is explored not just as an economic and political necessity but also as a valuable contribution to the cultural and social development of any society. Migrants and refugees are therefore not just to be seen as 'huddled masses' seeking twenty-first-century destinations but as potentially the most dynamic elements in societies that are seeking to advance in all respects. Thus, the book celebrates migration as something that is capable of enriching host communities economically and culturally. But to achieve this end migrants and refugees need guarantees to work, welfare and security. These are all too often denied to them, as the contributors to the following chapters demonstrate.

The chapters thus address some key problems faced by refugees and recent migrants in accessing employment that matches their skills and qualifications. Refugee and migrant populations in developed Western economies generally have higher levels of qualification and professional skills than indigenous workers but nevertheless experience higher levels of labour market exclusion or underemployment. This underutilisation of their skills has negative consequences for host economies, as well as for individual refugees and migrants. Thus, within the discussion of policy and labour market issues, the contributors collectively demonstrate the commonalities of experiences between the two groups, particularly in relation to their ability to access postmigration employment.

In an era when the number of refugees is growing worldwide and when, as an international phenomenon, movements of people are often directly linked to their search for work, the book addresses a number of key themes, exploring:

- State policies in relation to recent migrants and to refugees;
- The extent to which there are specific barriers that refugees or recent migrants face in accessing work;
- Whether these barriers are fixed, can be negotiated around or are constructed specifically to exclude refugees or recent migrants; and
- The extent to which refugees and recent migrants are able to develop their own strategies, in negotiating pathways into employment.

The collection contributes to the development of both a theoretical and empirical analysis of employment models and explores the extent to which they operate to include or exclude refugees and recent migrants from local labour markets. The book focuses specifically on employment and generally challenges existing orthodoxies, which have sought to explain the low labour market integration of refugees and migrants, mainly as a product of individual limitations—for example, lack of linguistic competence or of appropriate qualifications, rather than of structural discrimination.

The idea for this book came at the end of a thirty-two-month European Social Funded project on the labour market barriers faced by refugees. The project had explored the reasons why refugees experience particularly poor labour market outcomes in relation to their search for suitable work. Our findings differed somewhat from those of previous research as we found that linguistic competence could actually be a barrier to employment where the only available work was low skilled and low paid. We also found that the way that a person spoke the host language was as significant as whether or not they could speak it. However, while some of the following chapters are based on this research, the scope of the book is wider, with contributions from a wider range of research experiences, to ensure that the book could present an international picture of twenty-first-century migration. The concept of a book that could explore labour market issues for refugees and recent migrants together was also stimulated by the consequences of the accession of the eight countries (A8) of Central and Eastern Europe[1] which joined the European Union in May 2004. Of the previous EU15 countries,[2] the UK, Ireland and Sweden were the only ones not to place any restrictions on A8 nationals who wanted to come and work and soon it was clear, particularly in relation to the UK and Ireland, that this wave of migration was a new phase, representing the largest movement of peoples for work within Europe since at least the end of the 1939–45 war. The data are still unclear to give the exact number of those who made the decision to move for work, but available estimates suggest that anything between one and two million people have moved in the four years since 2004. In that year I had also begun to work on two research projects exploring the experiences of these new migrants, and it was clear that the stories that we were hearing from them, their reasons for migrating and their experiences in work in the UK resonated with the stories that we had been uncovering in the refugee research. And thus the idea of drawing these two communities, which are often separately differentiated, together, to understand how the trajectories of refugees and migrants combine and separate, was reached.

The book's chapters are organised around three key themes:

1. Concepts and methodologies;
2. State policies in relation to migrants and refugees; and
3. Structural discrimination and strategies of response.

In the first three chapters a conceptual framework is developed, providing an exploration of the employment predicament of refugees and recent migrants and demonstrating their common experiences within host countries. The chapters explore key concepts and methodologies useful in researching the labour market positions of refugees and recent migrants. The second of the themes is developed in Chapters 4 to 9, which provide an analysis of employment policies in relation to refugees and new migrants in Australia, Canada, France, Italy, the UK and the USA. In analysing these countries' experiences of recent migration, each of the chapters also focuses on one or more specific themes. These are legal regulation (the USA), visible difference (Australia), labour market demand (Italy and the UK), gender and migration (Canada) and racism (France and the USA). The third theme is developed through the contributions in Chapters 10 to 13, which examine structural discrimination in relation to employment and the extent to which it is possible to construct alternative strategies or specific support mechanisms that could ease entry into the labour market for refugees and recent migrants, including: whether employability initiatives may deliver positive employment outcomes for refugees; pathways into work and the extent to which refugees and migrants construct their own pathways.

Chapter 1 conceptualises the commonalities of experience between refugees and recent migrants setting the context for the later contributions, by theorising on these commonalities. The chapter does not argue that there are no differences between refugees and recent migrants and that immigration status and the reasons for migration are not important in shaping the experiences of workers who leave their country of origin, but rather argues that their labour market experiences may be more analogous than might be recognised and that, in particular, the experiences of undocumented or semidocumented migrants have strong echoes with the experiences of those seeking or who have sought asylum. I suggest that the labels of 'refugee' and 'migrant' are framed within concepts of state policies and of time and that at some juncture individuals should no longer be conceptualised as migrants or refugees but instead must become part of an indigenous minority ethnic community, to which labour markets may respond differently and for whom some of the constraints on labour market access—their lack of knowledge of how such markets operate; their fragile networks and so on—are no longer fundamental to their employment situation, while not negating the fact that they may still experience discrimination and disadvantage.

In Chapter 2 Allan Williams shows how academic and policy under-standings of migrant employment tend to be fragmented—by level of analy-sis, by discipline, and by type of mobility. Williams explores the concept of employability, which has sharply different meanings for individual migrants, employers and states. He advances three main arguments. First, the need to understand the extent to which individuals', employers' and states' under-standings of employability are mutually constituted but also inconsistent. Second, that increasing emphases on 'flexibility' and particular forms of skills and knowledge are changing how different actors understand employ-ment and employability. And, third, that employability remains the key to understanding whether migrants' first, and often suboptimal, labour market entry is a precursor to 'entrapment' or occupational stepping-stones. While demonstrating the complexities of this fragmented intellectual landscape, Williams also provides a useful starting point for the later chapters.

In Chapter 3 Paula Snyder and myself document the methodologies employed and the methodological challenges encountered in conduct-ing research among what are frequently conceptualised as 'hard-to-reach' groups and offer strategies, developed in the process of our refugee project, which may assist in overcoming such challenges. The chapter discusses issues such as how researchers can maintain long-term commitment from research subjects when engaged in longitudinal research, as well as documenting the use of visual methods to explore and disseminate research and to encourage the participation of research subjects.

Tessa Wright and myself, in Chapter 4, examine both the international and the UK national legal framework regulating the employment of refugees and of recent migrants. We focus on an analysis of the short- and long-term implications of excluding asylum seekers from the labour market and on the impact of successive and increasingly restrictive changes in immigration and asylum laws in the UK on the employability prospects of refugees and recent migrants, in particular where legislation constrains the right to work.

In the next chapter (Chapter 5) Val Colic-Peisker explores the intersec-tion of the labour market experience of permanently protected humanitarian entrants who arrived in Australia in 1990s–2000s and the Australian immi-gration and settlement policies pertinent to their experiences. Using data from a three-year research project, she focuses on the employment outcomes of three refugee groups in relation to their racial and cultural visibility and therefore the potential for discrimination in the labour market. The surveyed sample had high human-capital profile and relatively high English-language proficiency, but the employment outcomes, after an average of five to seven years of residence, were poor, which Colic-Peisker states is consistent with Australian government statistics showing that humanitarian entrants have considerably poorer employment outcomes than other migrant streams. Her research shows that the explanation is not to be found in their low human capital but rather that it is other factors that may be at work, principally labour market segmentation, which allocates the visibly different refugees into undesirable jobs. Colic-Peisker argues that labour market segmentation is achieved through systemic discrimination, reinforced by mainstream employer prejudices and the negative stereotyping of racially and culturally different immigrants and refugees.

Migration into Italy provides the backdrop for the next chapter, by Giovanni Mottura and Matteo Rinaldini. Here too data are presented to demonstrate how the position of migrants in the Italian labour market can be explained in relation not only to the specific characteristics of the migrants themselves but above all as to how these characteristics interact with the legal system and the internal dynamics of labour demand. Mottura and Rinaldini explore these themes by briefly examining the initial two periods of migration to Italy, in the 1980s and the 1990s, and then focusing on the current period, which they describe as marked by an increased presence of migrants within core sectors of the Italian economy. Labour migration is theorised through the concept of a 'reserve army of labour' in which migration becomes a core or key element in the processes of the reproduction of capital. In this interpretation, migration is seen as a factor in the increasing complexity of the theory, consequent on the fragmentation of collective bargaining, the creation of enterprises based on flexible organisational models, the dismantling of welfare and the delocalisation of manufacturing.

Nandita Sharma, in Chapter 7, shows how citizenship and restrictive rights to its access encourage undocumented status where labour exploitation is rife. Sharma argues forcefully that these outcomes are not accidental but are a consequence of state policies on migration. As she points out, where one migrates from, how that location is situated in the broader geopolitical concerns of the U.S., as well as how that location fits into the broader global capitalist economy; how one's formal educational qualifications are judged within the U.S.; how one is racialised and gendered; and, last but by no means least, how one is classified and positioned by the state in its hierarchy of differential statuses (citizen, permanent resident, refugee, temporary 'guest worker' or 'illegal') all have enormous consequences for the labour market experiences of migrant workers.

Steve Jefferys, in his chapter on France, provides a historical backdrop to migration into France, pointing out the role that migration has played in French history over more than a century. Jefferys explores migration through his work on a research project that focused on the role of trade unions in France in mediating racism and on the extent to which trade-union practice can enhance the employment experiences of refugees and recent migrants. Jefferys shows that national trade-union policies more rarely filter down to local-level implementation and that among French indigenous workers and their trade-union representatives there is an unwillingness to consider the adoption of different agendas to serve the interests of migrant workers. In the context of societal norms, based on the view that French citizenship brings with it equality of rights and equality of treatment, these encourage the rejection of any special measures to respond to the specific needs of migrant workers. Val Preston and Silvia D'Addario, in Chapter 9, focus specifically on the labour market experiences of migrant women. They demonstrate that the weight placed on education and skills, as criteria for admission to Canada, has not translated into economic success for recent newcomers and that migrant women are encountering serious and persistent challenges in finding employment commensurate with their qualifications and experience. Migrant women have higher unemployment rates and lower earnings than Canadian-born workers. Using data from the Gender and Work Database, they compare the economic circumstances of skilled migrant women who differ in terms of their marital status, racial identities, fluency in Canada's official languages, and period of arrival.

Alice Bloch, in Chapter 10, points out that while employment is a major part of the UK government's refugee integration strategy, current initiatives to help refugees into the labour market focus largely on individual employability and capacity building, rather than on discrimination or on reversing restrictive policies. Drawing on a large-scale survey, which Bloch carried out for the UK Department of Work and Pensions, she explores the interaction of personal characteristics, government policy and human capital with labour market outcomes. The chapter looks at existing 'barriers' to the labour market and at the difficulties refugees face accessing jobs that are commensurate with premigration skills, qualifications and work experience. In the subsequent chapter, Anne Green also looks at key issues that should be addressed to ensure that the economic contribution of refugees is maximised. Green highlights both the barriers faced by migrants and refugees in accessing and retaining employment commensurate with their skills and also looks at the employer response. She assesses the scope, nature, strengths, weaknesses and gaps of local initiatives to help integrate migrants into the labour market. More generally, the chapter discusses some of the generic challenges and associated dilemmas to labour market integration at the local level. These include a proliferation of local actors and problems of coordination between stakeholders, limited resources and a grants-based culture of provision, and an emphasis on short-term impacts rather than longer-term change and issues of bridging the gap between labour supply and demand—including the relative emphasis placed on employment (i.e. 'work first') vis-à-vis training. In Chapter 12, I look at how refugees and recent migrants access employment. I argue that while the UK government promotes open recruitment and also encourages employers to adopt transparent methods, the reality is that for many refugees and recent migrants informal methods of job access are more productive. The chapter also demonstrates that despite it not appearing so, many informal methods of job search are in fact highly formalised structures with their own rules of conduct. At the same time, while success, if measured by access to any job, is high through informal methods, such employment is more precarious and also contributes to longer term difficulties in securing employment commensurate with skills and qualifications.

In Chapter 13, Jenny Phillimore brings a comparative viewpoint to the subject, looking at employability initiatives for refugees in the European Union. Her chapter draws upon research undertaken to examine the range of employability initiatives being delivered across a number of EU states. She explores the initiatives on offer, the range of organisations that are providing them, together with the barriers to provision and good practice within existing provision. Phillimore finds that a combination of well-structured work experience, combined with positive employer participation and on-site accreditation of prior learning, is necessary for maximising refugees' chances of employment.

In the concluding chapter, I present an assessment of the future for refugees and recent migrants in the labour market. Based on the contributions in the previous chapters, I move towards an overall analysis of what the barriers are, whether these are real or 'imagined,' what routes there might be to successful labour market progress and also what the challenges are to it.

Sonia McKay
11 April 2008

NOTES

1. The A8 countries are: Czech Republic, Estonia, Hungary, Latvia, Lithuania, Poland, Slovakia and Slovenia.
2. The EU15 countries are: Austria, Belgium, Denmark, Finland, France, Germany, Greece, Ireland, Italy, Luxembourg, the Netherlands, Portugal, Spain, Sweden, and the United Kingdom.

Part I
Concepts and Methodologies

1 The Commonalities of Experience
Refugees and Recent Migrants

Sonia McKay

Within much of the literature on migration, the construction of a dichotomy between the position of refugees and economic migrants has been widely accepted. This is equally true in the context of legislative systems, where rights are differentiated between refugees, with recognised protection under international law, and economic migrants, subject to regulatory regimes that are conditional on: requirements for additional labour; the circumstances under which that labour is sourced; and the countries from which it is sought. That there should be such differentiation is generally accepted, since there are strong imperatives, particularly within the general context of increasingly restrictive state policies on migration that are premised on the ability to create a 'space' within the debate on migration, that allows those who 'genuinely' fear persecution to reach asylum.

However, while not wishing to ignore or minimise these differences, it is equally evident that the two groups share a number of common experiences, not only in relation to their reasons for migration but more importantly in terms of their labour market position within the host country that justifies them being discussed together as they are in this book. Thus, while the chapter will not suggest that refugees and economic migrants, in undertaking their migration journey, have identical imperatives, it demonstrates that there are sufficient similarities of experience to allow for them to be discussed together. The chapter also notes that while nominally differentiating between the two, certain government policy imperatives have encouraged the conflating of the two groups and the media and others have encouraged this, particularly within the public debates on migration. However, it does not follow that a critique of this policy agenda and media focus requires us to deny those commonalities. Rather, it forces us to argue why the categorisation of individuals either as refugees or as economic migrants forces individuals into fixed identities which do not really confirm to their experiences. The chapter therefore seeks to develop a conceptual framework around the background and the experiences of refugees and economic migrants, both in countries of origin and of destination, which allows us to explore the realities of their positions.

The chapter views both migrants and refugees as an integral part of the new migrant Diasporas, as it is increasingly difficult to make a clear distinction between 'voluntary' and 'involuntary' population movements (Crisp 1999). The chapter also does not distinguish between different types of migration (legal/illegal, temporary/permanent, forced/voluntary) as such distinctions do not express the reality of migration, as migrants cross the boundary from one category to another (Whitwell 2002). Salt et al. (2003) note how 'one type of migration or journey may transform into another' and that people's intentions change over time. My research has also concluded[1] that the initial aspirations of migrants and refugees for the future (which had usually envisaged a return home) can change over time, again indicating how migration itself contributes to new perspectives and ambitions, beyond those that might have been contemplated prior to departure.

DEFINING REFUGEES AND ECONOMIC MIGRANTS

The definition of a 'refugee', as laid down by the 1951 Convention relating to the Status of Refugees, is

> A person who has a well-founded fear of persecution because of his or her nationality, race or ethnic origin, political opinion, religion or social group; is outside the country of his or her nationality and owing to such fear is unable or unwilling to seek the protection of the authorities of his or her own country.

The term 'well-founded fear of persecution' is not separately defined within the convention, although there is a wide body of case law which has defined it. This is usually related to a threat to life or freedom (Gibney, 1988); however, the case law also makes it clear that economic oppression, where, for example, it is predicated on discrimination on racial, gender, or political grounds, may also fall within the definition. The International Convention on the Protection of the Rights of All Migrant Workers and Members of Their Families is the main legal instrument covering migrant workers, although it has been ratified by only twenty-seven countries and signed by a further fifteen, of which the UK is not one. Article 2 of the convention defines a migrant worker as: 'A person who is to be engaged, is engaged or has been engaged in a remunerated activity in a state of which he or she is not a national'.[2]

Thus, an examination of the wording of the two conventions suggests that the main difference identified is concerned with the individual's specific reasons for migration, rather than with the wider conditions which may act as 'push' factors in their countries of origin or as 'pull' factors towards their destination country. In reality there will be a wide set of circumstances leading to decisions to leave countries of origin. Whether individuals come as refugees or as economic migrants may relate as much to the legal systems they have to comply with as to their individual circumstances. For example, a worker from China may have experienced economic pressures to move but may also have experienced political repression. Whether that worker presents her/himself as a refugee or an economic migrant in the destination country will be as likely predicated on the policies of the host country towards the two different statuses as it would be on the individual's own situation.

Additionally we need to take account of the fact that even where such definitions are appropriate at the point in time when a refugee or migrant arrives in the destination country, we need to consider whether and, at what point in time, this 'label' is no longer appropriate. Is someone who arrives as an economic migrant still identified as such after thirty years' working in the destination country? Is a refugee always a refugee, even when she or he is permanently settled and has no plans to return, even where return is possible. In practice, these labels are more difficult to throw off where individuals are visibly different. Thus, while the Polish white worker, once he or she has acquired sufficient fluency in English is likely to be no longer identified as a migrant worker in the UK, for Black migrants the label of being a refugee or a migrant stays with them. Indeed, in many countries that label is applied not just to the generation that migrated but also to future generations, who are still described as 'immigrants' or the 'children of immigrants'. Thus, who refugees and migrants are and how they are seen is dependent on factors outside and beyond the legal definitions.

DIFFERENTIATING BETWEEN REFUGEES
AND MIGRANTS FOR POLITICAL ENDS

On 10 September 2007 UK Prime Minister Gordon Brown made a speech in which he offered an 'extra 500,000 British jobs for British workers'. The speech was delivered in the context of increased media attention on the number of recent migrants arriving in the UK, particularly since May 2004,[2] and was aimed at sending out a clear signal that government priorities were to prevent the arrival of new migrants who might be limiting job opportunities for resident UK workers. But the speech was also delivered in a context of limited evidence suggesting that migration had damaged job opportunities for indigenous workers, and indeed substantial contrary evidence pointing to migration having contributed to the UK's successful economic performance (Sriskandarajah et al., 2005). A report for the government's Migration Impact Forum, published just a month after the Brown speech, had found that recent migration had 'no discernable impact on unemployment' and had led only to a 'modest dampening of wage growth' for British workers at the bottom end of the wage league (Home Office, 2007).

Nevertheless, the UK government, along with the governments of most advanced industrial countries (as other chapters in this book make clear), proceeded with proposals to tighten migration further, by imposing a new and more restrictive immigration system, aimed at closing the doors to economic migration, particularly that which originates in sub-Saharan Africa, the Middle East, and Southeast Asia. The message is clear that Western governments will impose limits on migration for economic reasons and will also increasingly see economic migration as a temporary solution to labour-shortage problems and as a temporary route to work for global migrants. Why did Brown feel the need to show that the government would take a hard line on migration? Much of the answer relates to government policies, not on migration, but on asylum where, over the last two decades in most developed economies, there has been a strong policy position to restrict the number of asylum seekers. There were significant increases in the number arriving in the UK in the late 1990s, as a consequence of an increasingly dangerous world and due to the fact that journeys were less problematic, as routes between countries of origin and destination were relatively well established and existing communities in destination countries could provide both a 'magnet' for the newly arrived and advice and support to those planning to seek asylum. This meant that by 2002 the number of applications for asylum in the UK had reached 103,100. Of course this represented a very small proportion of the world's refugees, estimated by the United Nations at around 10.6 million in the same year. But a sustained and targeted policy by the UK government to limit the number of applications did result in a significant fall. By 2006 just 23,610 applications for asylum were made in the UK.

Thus, policies aimed at restricting asylum seekers, in the context of existing international protection, require of governments that they conflate the differences between refugees and migrants, so that an attack on economic migrants can be presented in the popular press as an attack on refugees, while governments can maintain that they are still complying with the Geneva Convention.[3] States are able to do this because they feed on an underlying lack of sympathy (and often hostility) towards persons deemed 'economic migrants' in the sense that they appear to have left their home countries 'voluntarily', merely to attain a 'better life' in the destination state, and therefore have no legitimate reason for seeking protection (Foster, 2007). Thus, the number of applications for asylum can be 'encouraged' to fall consequently conflating asylum and economic migrants as one group (Goodwin-Gill, 1996). Stevens (2003) shows how perceptions of the Roma as 'economic migrants' and 'street criminals' has led to the dismissal of many of their claims for asylum. Other examples include USA policy on Haitian refugees in the early 1980s (Farrer, 1995; Villiers, 1994) and on Vietnamese refugees by Hong Kong (Diller, 1998). It has also been used extensively in the media in Western refugee-receiving states, often as a justification in support of the call for 'tougher' measures in respect of asylum seekers.

Thus, the underlying distinction between forced or involuntary migrants responding to the 'push' factors of persecution (and thus deserving of protection) and voluntary migrants primarily influenced by the 'pull' factors of the attractions present in the receiving state (therefore undeserving of protection) needs to be challenged (Ghosh, 1998; Richmond, 1993). Choice is always exercised within specific constraints, and it becomes increasingly difficult to differentiate between, for example, workers who are obliged to work long hours for low pay because they have amassed large debts to enable them to migrate and who would face serious penalties from traffickers, agencies or other bodies if they did not work to pay, and those who 'choose' to leave their country of origin because they are political opponents or experience direct oppression, on the grounds of their beliefs or way of life. Distinctions are based primarily on two factors. First, the degree to which some individuals are seen as more worthy of sympathy in relation to their plight, while others are seen as essentially individualistic and acting to protect their own interests. This highlights the distinction that is made between those who are said to be exercising a 'choice' to migrate and those who have had no 'choice'. However, in reality choices may be extremely restricted in the case of economic migrants, nor is it the case that there are no choices in the case of refugees. As Harding (2000) notes:

> In the past, refugees have won greater international sympathy than economic migrants. Theirs has been the more identifiable grievance: at its source there is often an identifiable persecutor. Yet the order of economic difficulty that prevails in some parts of the world is akin to persecution. No consensus exists about the identity of the tormentor, and so those who try to put it behind them are more easily reviled than others fleeing the attentions of secret police or state militias.

In practice each individual makes assessments, based on their individual health and security, that of their family, friends and associates and also in relation to their own future economic security. But essentially, the factor that is most likely to affect individuals and their decisions on how to migrate, where to migrate to and under what basis, is the legal situation they will find. Refugees will more likely move either to countries nearest to them[4] or to countries where it is believed that the refugee policy agenda is most favourable to them.

Foster (2007), in her recent book, also questions the traditional distinctions between 'economic migrants' and 'political refugees' and suggests that, notwithstanding the dichotomy between 'economic migrants' and 'political refugees', the Refugee Convention is capable of accommodating a more complex analysis, which recognizes that many claims based on socioeconomic deprivation are indeed properly considered within the purview of the Refugee Convention (Foster, 2007). For Foster, as for Goodwin-Gill (1996) and Anker (2002), it may be increasingly difficult to distinguish between those whose flight is from poverty, economic degradation, and disadvantage and those whose flight is more generally accepted as legitimate when fleeing political or other oppressive situations.

Research shows that, in relation to both international migrants and to refugees, they do not primarily come from poor, isolated places but from regions and nations that are undergoing rapid change and development (Massey, 1998: 277). Their presence can contribute to the resolution of skills and labour shortages. Additionally the fiscal contribution of the foreign-born population to public finances is growing. Between 1999 and 2004 foreign-born workers in the UK had become proportionately greater net contributors and 'far from being a drain on the public purse, immigrants actually contribute more than their share fiscally' (Sriskandarajah et al., 2005). However, while there has been a significant amount of research on the economic contribution of foreign-born workers, there is relatively little research about the characteristics and motivations of different migrants and refugees and of their economic and social impacts and experience (Glover et al., 2001: 49). There is equally little research on the reasons why international migrants leave their country of origin and what their hopes are for life in their new country, regardless of whether and to what extent they were able to exercise a choice over when they left and where they arrived. Research on migration into Australia[5] has identified 'pull' factors for migration as: proximity, cultural tolerance, political freedom, higher wages, security, better social safety nets, better working conditions and more stimulating/challenging work. As far as refugees are concerned, a UK Home Office study (Robinson and Segrott, 2002) based on interviews with sixty-five asylum seekers has looked at their values, attitudes, and expectations on arrival. This found that key factors in choice of destination were: belief that the UK was a safe, tolerant, and democratic country; previous links between their own country and the UK, including colonialism; and the ability to speak English or the desire to learn it. The presence of family and friends was also a factor where the asylum seeker was able to exercise a choice of destination.

This existing body of research thus makes it clear that there are some areas of common experience, making it appropriate to look at international migrants and refugees, not as two separate groups with entirely different experiences but as two groups occupying different spaces but within a single trajectory.

CONVERGING AND DIVERGING EXPERIENCES

The position that refugees and recent migrants find themselves in regarding access to the labour market and use of job search methods is not identical, given that they may be subject to different legal controls and rights to work. However, there are factors that mean that in practice refugees and recent migrants often end up working side by side in similar jobs. It is for this reason that it is useful to develop a conceptual framework that provides explanations both for the constraints and possibilities that govern decisions to flee or to migrate, usually characterised as predominantly constraints in 'push' countries and possibilities in 'pull' countries.[6] Such a typology aims to locate individuals, both as refugees and as economic migrants within an overall map of constraints and possibilities. Table 1.1 sets these out.

The factors listed in Table 1.1 demonstrate that there is a continuum from negative to positive which shapes decisions to leave one's country of origin. The first factor relates to limited mobility. The fact that an individual may not be able to move either geographically or professionally within their country of origin is a likely factor in any decision to leave. For those fleeing persecution, limited mobility in country of origin is almost always present. Individuals may have been imprisoned, subject to house arrest or been restricted in their movements within the country, for example, as a result of ethnic conflict. Professional mobility may similarly have been restricted or limited. Political opponents of the established order are also likely to have had limited or no opportunities to follow their chosen profession. But geographical and professional mobility may also have been limited in the case of economic migrants. A lack of financial means restricts individuals' abilities to exercise their freedoms to move within their countries of origin. Similarly, a lack of jobs limits the individual's chances of exercising professional mobility.

A lack or absence of political freedom or the presence of oppression or dispossession is more likely to represent a push factor in relation to refugees, but again it is not only refugees who are affected by these factors. Although it is asylum seekers who are most likely to have fled for reasons connected with oppression, those identified, as 'economic' migrants may still have been motivated to leave their country of origin for reasons connected

Table 1.1 Factors Shaping Decisions to Migrate

	Experience in Country of Origin	*Perception of Country of Destination*
Negative	Limited mobility either for economic or political reasons	Proximity to country of origin
	Lack of political freedoms/ oppression/dispossession	
	Lack of jobs	
	Lack of accommodation	
	Low pay	Availability of jobs
	Restricted social opportunities	
	Possibility of successful return	Higher wages
	No possibility of return at least in the foreseeable future	Accessibility of opportunities
		Legal rights
Positive		Possible successful settlement

with political or military oppression. A regime of oppression may itself lead to a lack of employment, to poverty and to insecurity and may therefore represent a significant push factor for economic migrants, just as it would for refugees. A lack of work and lack of accommodation or the predominance of low pay all represent strong push factors for economic migrants. But refugees, particularly those of limited economic means, also are affected by these push factors. Whether an individual can survive in a situation where their lives are at risk is to some extent dependent on their economic means, which may enable them to buy protection. An individual's economic or political position may also influence his or her ability to engage in wider social networks and to exercise the right to a social life. Those fearing persecution are obliged to limit their social networks to those whose positions they share. Those whose economic means are restricted are similarly likely to have access to restricted social networks.

The last of the factors identified within the overall typology of country of origin 'push' factors relates to the predictability, at time of departure, of whether a successful return is likely or not. For both refugees and economic migrants, thoughts on departure are closely based on the notion of return. For the economic migrant return is normally seen as the consequence of successful migration; for the refugee return is hoped for when the conditions that led to departure have been removed. Migration is more rarely predicated on a notion of no opportunity of successful return, yet the reality, for migrants and for refugees, over many generations is that their journey is essentially 'one-way'. While those who successfully make the transition may re-visit their country of origin, permanent return becomes less likely the longer the period of migration.

THE EXERCISE OF 'CHOICE'

Much of the current debate within the UK, on refugees in particular and on migrant workers in general, has focused on an assumption that they choose to come to the UK and do so because they perceive that it offers good public services and 'generous' state benefits. However, the limited data available disputes this. For example, Robinson and Segrott's study showed that most of their respondent asylum seekers had very limited knowledge of what financial support they would be entitled to (Robinson and Segrott, 2002). Research carried out for the Scottish Executive, based on a survey of 523 refugees and asylum seekers, also found that many of those surveyed wanted to work and disliked being reliant on state benefits. Those unable to work often felt unhappy, frustrated and bored (Scottish Executive, 2005).

Choices about destination are also influenced by the legal rights that are believed to be present in the destination country and the extent of legal protection they offer to workers in general and to migrants and refugees in particular. These protections may be seen in relation to the individual migrating or to a wider group—for example, their family or a future generation. As Madood (2004) acknowledges, the desire to better prospects for their children is one economic motivation of migration, and our research has confirmed that family commitments affected both how individuals arranged their personal lives and how long they planned to stay. Crisp (1999) also notes that transnational social networks play a role in the decisions on where to seek asylum, particularly by providing information about the quality of life in the new country. Our research similarly found that of those refugees in a position to exercise a choice over their destination, the decision to choose the UK had been heavily influenced by the preexistence of family networks.

The conceptual framework outlined previously also assumes that there are reasons that may push those considering migration towards certain destinations. But this does not imply that particular destinations are freely chosen. Rather, the concept of migration always involves a selection from a limited range of choices, while in some cases the 'selection' is made by others. In the typology, five 'pull' factors are identified that may contribute to how selection is made. The first of these relates to proximity to country of origin. In the case of refugees, most flee to the nearest safe border to where they reside. Similarly in the case of economic migrants, destination is strongly related to distance from home country. This is for reasons of cost, ease of journey, and because it apparently represents the best situation for eventual return. There is no doubt that the recent large-scale movements of workers from Central and Eastern Europe, particularly to the UK and Ireland, were facilitated not just due to the existence of a favourable legal environment for migration but also due to the fact that the systems of transport both facilitated and encouraged movement between country of origin and destination.

The availability of jobs, the existence of jobs at higher wages, or the potential for greater occupational opportunities (or at least a belief that this is the case) is a fundamental requirement in the decisions of economic migrants to move. But it is equally an important consideration for refugees. In their survey, Robinson and Segrott (2002) found that in the vast majority of cases, while employment did not play a dominant role in the decisions of asylum seekers to migrate, those who had specialist skills had believed they would be able to utilise them. Our research too has found that when asking refugee respondents to recall what their thoughts for the future had been when they first arrived, most had thought they would continue to be able to work in their previous professions or occupations, even though they knew there might be hurdles to overcome to achieve this. Particularly in the case of those who had the opportunity to plan their departure, the researchers found a strong desire to engage in paid work as soon as possible following their arrival (McKay et al., 2006).

Refugees may have more limited choices over where it is that they seek refuge, as their options are constrained, due both to the policies of potential host countries in relation to refugees and to the availability of routes of exit and entry. Relying often on third parties to arrange transportation, the refugee may have to accept whatever option is presented, and this is particularly the case where the individual has few financial resources. Our interviews with refugees reveal that for some the decision on where they should go rested with the agent responsible for transportation. In one case the individual had wanted to go to Turkey but, having been advised that it too was not safe for him, put the matter in the hands of the smuggler; and, as he noted, 'they chose this country and that is why I came here'. In another case the refugee's father had been in negotiations with an agent to get his son to safety in Germany, and that is where this young man thought he was heading. It was only after twenty to twenty-five days shut in a truck that the doors opened and the agent said, 'Get down'. He replied, 'Is this Germany' and was told, 'No, this is England'. Yet another refugee had been told he was going to Minnesota but had ended up in Birmingham. Another said, 'If it were my decision I would have gone to America or Canada, because in Canada I know a lot of people—this is not a matter of choice. It is a matter of escaping and everything was arranged back home so I was in the plane and finish'. Almost half of the refugee interview sample (McKay et al., 2006) had not exercised a positive choice over where they should flee. As one female refugee pointed out, 'Normally you don't choose'. Indeed, in her case, had she been able to choose where she could go, it would have been to the USA, where she had a sister, or even to the Lebanon, where she had a brother. Another of the refugees interviewed had not chosen the UK, although as it happened he thought it was 'the best country for me to come to, but it's not my choice'.

Our research on migrant workers (McKay et al., 2006a; McKay and Winkelmann-Gleed, 2005) did find that migrants' stated motivations for coming to work in the UK were primarily economic; however, there was often an accompanying range of factors that had led them to migrate to Britain. These might include a lack of opportunity to work at home—due to unemployment or the impact of age bars or other forms of discrimination—and a desire to start a new life somewhere else. Some migrants did believe that life in Britain would provide their families with better opportunities, in terms of health care and schooling or more stability, while others had a combination of economic and political reasons. Migrant workers interviewed gave varied reasons for coming to or choosing the UK. Common reasons were to earn more money, find work, experience new things, join existing family members, improve their English and as a stepping-stone to a third country. Reasons for choosing the UK included the fact that there were historic links between their home country and the UK and secondly that English is a universal language.

In the case of refugees, while the primary reason for leaving their country of origin was for sanctuary, the choice of where to flee to (in those cases where it has been possible to exercise a choice) was based on a variety of reasons, some of which were common to those expressed by international migrants. Thus, for both groups historic and family ties and knowledge of English (or a desire to learn the language) were important factors in their decisions whether on migration or flight. In addition, as the research shows, migrants' choices over where to go to find work are also limited by destination-country policies in relation to migration. And for those who as a consequence enter without documentation, the decision on where to go is more likely to be in the hands of the smuggler as in the hands of the migrant worker. Among the migrant workers interviewed, not all had favoured the UK as their destination of choice or may only have seen the UK as a stepping-stone towards an ultimate preferred destination. One respondent had wanted to go to Canada, or to the USA, because he had friends in South Carolina. In both research projects, where there had been the exercise of a choice, a belief that there were better opportunities was the most likely reason given by both migrants and refugees for choosing to come to the UK.

The last factor identified in the conceptual framework of 'pull' factors relates to possible successful settlement. For those who do not see a future in their country of origin, then an ability to settle in the destination country is a strong motivator for choice. However, rather than seeing this as an initial pull factor it is more correct to describe it as a pull factor which can only be more clearly visualised once migration has occurred rather than on arrival, other than in cases where the individual is joining an already established community which had already decided to settle permanently. In these cases the migrant or refugee will view this potential for successful settlement as a key pull factor.

In the research that we have undertaken it was not every case that respondents were confident about the decisions they had taken in relation to migration. Some were explicit about the disappointments they had faced. They were disillusioned, missing their families, had lost their jobs and found it difficult to establish a support network. Thus, not all experiences of migration were viewed positively. Particularly in the case of those migrants who were in a position to exercise a choice to return, some spoke of returning earlier than planned, mainly due to the dismal state they found themselves in. They spoke of being overwhelmed by the high cost of living or by the irregularity of their employment, a situation which left them with expenses but insufficient earnings to meet them, making it impossible to send money back home, despite that having been a motivation for coming to the UK. In some cases their current circumstances were almost too depressing to enable them to see a way out. One woman's description of her current life encapsulates the predicament of many of those interviewed: 'I also write poetry. I always kept a journal, since I was a child. But in England, I lost my inspiration for writing poetry. This country doesn't give me inspiration. I never wrote anymore'.

CONCLUSION

An individual's experiences and motivations for migration will vary according to the personal, political, economic and historic situation they find themselves in, and consequently there is no such thing as the 'typical' migrant worker or refugee. Without understanding the full background of where individuals have come from and what caused them to make the decisions they have made, one has to be careful not to treat migrants or refugees as two distinct groups but rather to see both as a consequence of an increasingly instable global system which forces individuals to leave their countries of origin and which gives them limited control over the core decisions about their futures. As this chapter has argued, it is important not to see the two groups as having different trajectories, interests, and experiences. Particularly in their early periods in the destination country, as the chapter has argued, their experiences of work are similar. Refugees and economic migrants work alongside one another, doing the same kind of low-paid work, which has often been rejected by indigenous workers. It is in these workplaces that refugees and migrants can find that their common experiences of migration may also be replicated by their common experiences postmigration.

NOTES

1. These conclusions are primarily based on three research projects, undertaken between 2004 and 2007. One project, funded by the HE-ESF, looked at the labour-market position of refugees. The others, funded by the East of England Development Agency and the Health and Safety Executive, respectively, focused on the work skills and health of recent migrant workers.
2. In May 2004 ten countries acceded to the EU. The UK, along with Sweden and Ireland, permitted the entry for work of citizens from these new EU member states and, although the exact numbers who did migrate are not known, from the available data it is possible to say that at least one million workers did move to the UK alone.
3. The 1951 Geneva Convention Relating the Status of Refugees as amended by the 1967 Protocol.
4. It should be remembered that the vast majority of the world's refugees never make it to the rich countries of the West. Instead they are reliant on neighbouring countries as their principal source of support on flight.
5. Joint Standing Committee on Migration (2004).
6. The concepts of 'push' and 'pull' are used here only in their most general sense and are not intended to convey the notion that migration can be neatly categorised as into 'pull' and 'push'.

2 Employability and International Migration
Theoretical Perspectives

Allan M. Williams

This chapter provides a review of the relationship between employment and migration, focussing particularly on how this is manifested in terms of employability. Employability is the 'ability to bring a particular kind of knowledge to a task, and be able to collaborate effectively with others to achieve a common task' (Bentley, 1998: 103). This has two aspects, which broadly equate to quantitative and qualitative dimensions. In 'quantitative' terms, employability can be understood to encompass the ability of migrants to secure jobs, to maintain these and to obtain other jobs, whether by choice or as a consequence of labour market restructuring. In contrast, in qualitative terms it can be understood to implicate the types of jobs obtained and how these relate to the skills, competences and knowledge of the migrants.

International migration can be seen as one of the means by which individuals respond to employability issues. They may migrate because there are better opportunities for securing employment in other countries, whether in terms of getting any job, the security of that job, or the returns measured in wages, working conditions, or the capacity for learning. They may also make decisions about employability in terms of differences between the country of origin and the destination. Or they make take the longer term 'spatial view' and consider how international migration—or more precisely the learning and knowledge opportunities this affords—will influence their employability in their country of origin, or some other country, at a later stage in the cycle of migration. In its most positive form, this interpretation posits international migration as being a source of exceptional learning, particularly of nationally specific encultured and embedded knowledge that allows individuals to enhance their employability over time. The migrants, in this reading, are reflexive individuals, for whom migration is a component of lifelong learning and internationalised careers.

The previously stated understanding of employability is based on bounded rationality and a substantial degree of voluntarism, indicating the significance of human agency in international mobility. This is, of course, a far cry from the experiences of many migrants. For refugees and asylum seekers—whether de facto or de jure—and for many labour migrants and their families, employability issues are first and foremost about rights to seek employment in the formal economy. Beyond this, employability is about securing jobs with 'fair' wages, employment and working conditions, at least in comparison to indigenous workers, if not in terms of broader distributional issues implied by competing notions of social justice. But employability is also about their ability to retain these jobs and/or secure other jobs which offer greater personal returns, however measured. In other words, employability is also about learning and the social recognition of migrants' knowledge. This puts employability in the context not only of workplace learning but also of broader structural features such as the changing social divisions of labour, regulation and institutions, and issues of social identities, social recognition and discrimination. It is the intersection of these mediating influences that determines whether international mobility leads to labour market entrapment or stepping-stones for individual migrants.

This chapter aims to set out a conceptual framework for analysing the relationships between employability and international migration. It examines the following themes: the significance of changing forms of international migration; the different meanings of employability to individuals, firms and national states; the changing nature of employability; and outcomes in terms of labour market entrapment versus stepping-stones.

CHANGING FORMS OF INTERNATIONAL MIGRATION AND EMPLOYABILITY

The diversity of individual experiences of employability can be approached through a number of cross-cutting migration typologies within which individuals can be located. These shape how individuals experience employability.

First, *the intersection of migration and employment regulations* determines access to the formal and informal labour markets. On the one hand, this involves the broad approaches of the state to employment and welfare, as typified by the 'varieties-of-capitalism' argument (Hall and Soskice, 2001). While neoliberal tendencies have led to some convergence across Europe (Esping-Anderson, 1996), the national state remains the key site for employment and welfare policies (Favell and Hansen, 2002; Hudson and Williams, 1999). For example, the relatively small numbers of nonregistered migrants in Scandinavian labour markets are related to the tightly regulated nature of those markets (Hjarno, 2003). National migration policies are to some extent influenced by these broad differences in varieties of capitalism, but are also driven by the politics of migration. As Morris (2002: 410) comments, migration policy 'can be understood as the "management of contradiction", in which policy and practice seek to strike a balance between concern over national resources, which tends to limit entry, and continuing employer demand and the assertion of human rights, which potentially expand entry'.

Migration and employment policies constitute what Brubaker (1992: 23) terms 'both an instrument and an object of social closure'. Immigration and employment laws tend to be implicitly gendered, ageist and racist in the way that they close or open particular labour market segments to particular migrant groups, with consequences for employability. They are also increasingly skill- and sector-specific, privileging so-called highly skilled migrants, although there are significant differences in the specific policies deployed by national states, whether in relation to labour migrants or refugees (McLaughlan and Salt, 2002). The result of the prioritisation of skilled labour has been that the demands of employers for less skilled workers have been met, in many and perhaps most countries, by clandestine migrants and refugees or asylum seekers, rather than by labour migrants. The outcome, however, is that 'Neither the more "flexible", pro-globalisation regimes of the UK and Ireland, nor the more social protectionist regimes of France, Germany, the Netherlands or Denmark, have been able or willing to do much about this, while the governments of Southern Europe have been quite unprepared for it' (Jordan et al., 2003b: 197–8). In reality, there has often been a fluid situation, with migrants moving between compliance, semicompliance and noncompliance (Ruhs and Anderson, 2006), seeking to enhance their employability in response to changes in their personal circumstances, the economy and the regulatory framework. This reflects the way in which different regulatory regimes—for example, for employment and immigration—intersect in often a 'contradictory and adhocratic' manner (Guiraudon, 2003).

Secondly, there are obvious differences in employability, in terms of whether labour migration is realized via *intracompany mobility versus 'free agent movers'* (Arthur and Rousseau, 1996; Williams, 2006). Migrants moving within companies have guaranteed jobs, of course, so that employability for them is a qualitative issue of how this international experience influences their long-term employability (McCall, 1997). In contrast, for international free-agent movers (the term 'free' is conditional, of course) mobility has to be understood in terms of their employability in their countries of origin and prospects of securing any, or particular types of jobs, in the country of destination.

Thirdly, *the nature of migration* is changing. The classic (but overstated) view of the once-in-a-lifetime migrant who eventually becomes a return migrant (King, 1986) has been supplemented by other models (Hardill, 2004; King, 2002), including nomads, whose travels could take them anywhere (Clifford, 1997), and serial migrants (Ossman, 2004) involved in cycles of migration to single or multiple destinations. Migration has also tended to become more short term—although this applies more to migrants from the more developed than the less developed countries. Refugees and nonregular migrants also face more constraints than regular labour migrants in moving across international borders, although there is evidence that their position is sometimes more fluid than is assumed (Jordan and Duvell, 2002). Returned or circular migration can be a response to migrants' changing employability (loosing jobs or facing lower wages) but also determines their employability, particularly in companies or industries where promotion and advancement are determined by length of services.

Fourthly, there are also significant differences in the employability of primary *or lead migrants as opposed to trailing migrants, whether spouses, children or elderly relatives.* Immigration regulations frequently restrict the employment rights of accompanying migrants, perhaps absolutely, or in terms of the numbers of hours they can work. The result is that employability means different things to lead and trailing migrants, and these differences are gendered and age related (Kofman, 2004). It is therefore important to theorise the contrasting employment, career and employability experiences of migrants within households. This does not imply that all 'accompanying persons' face reduced employability prospects and are necessarily passive victims of regulatory regimes. Some national states have liberalized the employment rights of the spouses of skilled migrants in order to provide a more attractive regime for the latter in context of global talent wars. In other cases, migrants excluded from formal employment have sometimes turned to various forms of self-employment. But the resources that this requires also underline the mediating influence of social class on migrant employability.

The employability of migrants and refugees is determined then by their positions in relation to a complex array of immigration and employment regulations, which vary between countries. Moreover, these regulations change over time, as do the rights to permanent residence or citizenship. The result is that employability is the outcomes of 'a bricolage of territories with differentiated rights for different migrant groups' (Williams, 2001: 103).

MULTILEVEL UNDERSTANDINGS OF EMPLOYABILITY

Although this chapter started with a single definition of employability, this has different meanings for different social actors. In particular, individual migrants, employers and the national state understand employability differently. Of course, there is no shared understanding within any of these three categories, but here we focus on the differences amongst them.

The National Level In economic terms, national states have to take a fundamental decision as to how to accumulate stocks of human capital, whether via indigenous sources (that is, via investment in training and education) or external ones (that is, via immigration). In practice there are a number of discourses that influence immigration policies, ranging from alleviating labour shortages or supporting national growth and productivity strategies to promoting postcolonial or other political relationships and facilitating cultural ties and exchanges (Ruhs, 2005). However, policies increasingly are influenced by neoliberal discourses (Jordan and Düvell, 2002), with significant implications for how the employability of migrants is understood at this level.

In general, there has been a broad shift in the European Union and in individual member states towards understanding the employability of migrants in terms of filling specific labour skills shortages (notably in the IT and health sectors) and in terms of making national economies more competitive (Jordan et al., 2003a), including addressing the issue of flexibility (Green and Turok, 2000: 599). The shift is far from uniform, of course, and has favoured more skilled workers, as well as particular sectors which face labour shortages. This resonates with debates about the knowledge-based economy and is necessarily elitist in terms of occupations, sectors and places (for example, skilled immigration policies in the UK are to a considerable extent driven by the specific needs of London as a global city). In contrast, immigration policies for less skilled workers are usually framed in terms of meeting supposedly 'short-term' labour shortages and are typified by particular regimes for seasonal agricultural workers, construction, or care workers. In reality policy shifts are uneven and discontinuous, being implemented within migration systems that for several decades have been seen as national custodians of a Fortress Europe (Geddes, 2000). Policies towards refugees and asylum seekers are driven less by such economic logic and more by the interface between the domestic politics of immigration and political visions of countries' roles in the international order. Employability features little in the discourses about such mobility, other than in relation to employment rights.

The Firm/Employer Level Employability for individual firms is understood in terms of their production strategies and increasingly—although not exclusively—this has focussed on the notion of firm-level flexibility in context of neo-Fordism (Amin, 1994). Labour market flexibility lies at the heart of flexible production strategies, and there is a well-developed literature in this area. In practice, there are a number of different discourses surrounding flexible labour in firms, which ascribe different roles to the flexible worker: they can be 'flexperts' (van der Heijden, 2002) who rapidly acquire expertise, sources of just-in-time expertise (Brandenburg and Ellinger, 2003), 'warm bodies' for hard-to-fill vacancies, or gurus who transfer critical knowledge to firms (Barley and Kunda, 2004). There is also a well-established literature on generic labour market flexibility dating back to Atkinson (1984) and later researchers in the same vein (for example, Regini, 2000). They recognise four types of labour market flexibility—numerical, functional, temporal and wage. Each of these relates to different notions of employability in the employer's perspective—ease of recruitment/dismissal, amenable to moving between jobs within the organization, willing and able to work variable hours in response to fluctuations in demand or production conditions, and an aggregate ability to manage wage levels.

There are many ways in which employers can source the labour required by these different flexibility strategies, and labour market segmentation is often at the heart of these, with gender, age and ethnicity commonly being implicated. However, international migration is increasingly significant for the flexibility strategies of firms in many countries. This can be illustrated by the following idealised schema (Table 2.1), which illustrates some of the key issues.

The employability of migrants is of course viewed differently by firms in different economic sectors and countries. For example, in Silicon Valley Chinese scientists are a source of substitute human capital in the face of declining numbers of American PhD science students, but in China the same individuals as return migrants are an important source of embedded knowledge and social networks (Saxenian, 2006). In terms of filling less skilled posts, some firms prefer employing unregistered migrants because of their greater vulnerability and consequently being more amenable to flexible employment practices; this can also be a way of creating competitive relations between unregistered and registered migrants and between newly arrived and more settled migrants (Solé and Parella, 2003: 129). Similarly, Matthews and Ruhs (2007) report on the (flexibility) advantages of employing unregistered migrants for employers in the hospitality industry.

The Individual Migrant Level Individual migrants are neither economic dupes nor unconstrained human agents (Hudson, 2004); hence, their employability should be viewed as shaped by, and shaping, the firm and national levels. For many migrants employability is about survival, or about fulfilling specific economic goals in terms of savings or remittances, rather than about learning and career progression. Indeed, the career implications may be negative, especially where length of service is a condition of promotion. Trailing spouses are particularly likely to incur the career penalty of moving with a partner who is the lead migrant.

Table 2.1 Idealised Relationships between Flexibility, Employability and Migration

Type of Flexibility	Employability Association	Implications for Migration
Numerical	Casual and seasonal employment	Work permits, visas
Functional	Multitasking	Assumed attitudinal competences
Wage	Low-wage acceptability	Vulnerable and informed by lower wages in countries of origin
Temporal	Flexible working hours	Seasonal work permits. Limited alternatives. Short-term economic goals.

Source: Author.

However, all migrants are reflexive and to varying degrees can utilise regulations and firm practices to their advantage—up to a point. Hence, migration and flexibility are increasingly seen as something to be aspired to (Ong, 1999) and as contributing to career development. This is associated with a number of theoretical perspectives, including the Foucaldian view of the individual as a self-governing entity (see Hardill, 2004, on migrants), and Beck and Beck-Gernsheim's (2002) notion of individualization (see King and Ruiz Gelices, 2003, on student migrants). Moreover, the meaning of migration in terms of employability for individuals is not fixed—rather, it may shift at various points during the migration cycle. For example, after first migrating, individuals may be less employable because they lack nationally specific human capital, but then their employability may increase as they acquire such knowledge (Becker, 1975). If they decide to return to their home countries, they may find that their employability is even greater if their sojourn abroad has resulted in the acquisition of particularly scarce human capital (Dustmann and Weiss, 2007).

While we have discussed these three levels separately, they are in fact interrelated. This is similar to Amin's (2002) notion that different scales are folded together—what happens at one level is both influenced by and influences other levels. Or as Goss and Lindquist (1995) stated, specifically in relation to migration, this can be understood in terms of how life cycles and individual careers intersect with broader changes in the organization and structure of the economy.

THE CHANGING NATURE OF EMPLOYABILITY

Structural economic changes have been accompanied by a relative shift to more discontinuous careers for individuals. This has meant a shift from employment security to employability security (Opengart and Short, 2002: 221)—in other words, the expectation of being able to find new employers in a labour market where employment is understood to have become increasingly uncertain. Gold and Fraser (2002) argue that there has been a shift away from seeing careers as planned linear progressions within organizations, with strong internal labour markets, and long-term employment with individual employers. Instead, individuals increasingly have 'boundaryless' careers, which include 'a range of possible forms that defies traditional employment assumptions' (Arthur and Rousseau, 1996: 3).

This has important implications for understanding the employability of many individual labour migrants, that is, their ability to find new employers in the course of their 'boundaryless' careers. In its most limited sense, this simply means that migration is a way of extending the options that are available when confronted by labour market uncertainty. However, for selective migrants who are more favourably positioned in the labour market, migration may be empowering in terms of advancing their careers. For these more favoured migrants, the migration experience can be constructed around the acquisition of what Sennett (2000) terms 'flexpertise'—the ability to learn and adapt quickly to changing circumstances. In other words, migration can be about acquiring learning competencies, or learning how to learn. Migration may be interwoven into the flexpertise that allows boundaryless careers to be pursued, while enhancing individual employability—although these relationships are mutually interdependent.

For refugees and many labour migrants, who previously had no employment security at all, the shift from employment security to employability security is probably of limited relevance compared to the overriding need to secure their economic survival and basic rights of residence or citizenship under adverse circumstances, at least in the initial stages of migration. For them, discontinuous employment is not being transformed into boundaryless careers characterised by continuous learning to learn but by increasing labour market insecurity and impoverished employment opportunities. This is explained by Solé and Parella in relation to those migrants who are also ethnic minorities subject to racism (2003: 124):

> Racism as a system of the 'ethnification' of the labour force permits the increase or decrease, according to the needs at any given time, of the number of individuals available for those economic activities that are worst paid and least attractive in a given moment of time and space.

The determinants of changes in employability for them are far more likely to be rooted in the politics of discrimination and racism and how these are contested in both broader societal struggles and by the migrants themselves via both collective and individual actions. The employability of all migrants, irrespective of ethnicity, will also be influenced by state and voluntarist initiatives to enhance their employability (such as language classes and other forms of training, as well as social welfare and housing policies).

MIGRATION: LABOUR MARKET ENTRAPMENT OR STEPPING-STONES?

We have already noted that migrants, because their knowledge is not fully recognised (perhaps because they lack actual, or designed-to-be exclusionary, national knowledge), or because of obstacles such as labour market regulations, may enter the labour market suboptimally. This is fundamental to the thinking behind human-capital theories (Becker, 1975; Chiswick, 1978). However, it does not automatically follow that the acquisition of nationally specific human capital over time will lead to higher wages and occupational mobility for the migrants. The eventual employment outcomes depend on whether these initial jobs constitute labour market stepping-stones or labour market entrapments (discussed further in Williams and Baláž, 2008).

In the case of 'stepping-stones', there is a gradual matching of knowledge and occupational position as migrants overcome barriers to using and acquiring knowledge. Migrants use their initial jobs as stepping-stones to jobs that are more rewarding in some way. To some extent this is based on positive assumptions about the employability of migrants, although this is also informed by sectoral and individual differences. In the case of 'entrapment', an initial suboptimal labour market entry has enduring consequences, as individuals become 'trapped' in a particular job or labour market segment, which implicitly means not only a failure to progress in their careers but also being condemned to relatively poor returns, whether in terms of wages or learning.

It is, however, necessary to place this evaluation in context of the cycle or cycles of individual migration. A migrant may move to a job abroad in which (s)he appears to be entrapped, not being able to make further progress in the destination-country labour market. However, their experiences in this job may provide stepping-stones to jobs in the country of origin. For example, individual migrants may accept suboptimal jobs abroad in order to acquire particular knowledge for which a premium is paid in their countries of origin; this may be a technical skill which is in abundant supply in the destination and for which a low return is payable compared to the country of origin, where there is a specific skills shortage. This is a classic illustration of the notion that migration may constitute 'significant learning moments' for individuals, whereby they acquire particular forms of knowledge and enhance their employability—although only in particular places.

The occurrence of such significant learning moments is especially pertinent if we look beyond formal qualifications and technical knowledge to the acquisition of a range of social skills and competences, such as self-confidence, networking skills, learning and adaptability competences and self-reliance (Williams and Baláž, 2005). It may also be the case that employment in a job that seemingly represents labour market entrapment can be construed as a stepping-stone because it provides a platform for learning outside of the workplace—of, for example, social or language skills. This is illustrated by the example of au pairs working in the UK, a caring and cleaning job which in itself usually is not a stepping-stone. However, migrant au pairs do acquire social skills and competences, as well as language skills, both in and outside of their work roles, for which a premium may be payable in their country-of-origin labour market. In this instance, these jobs do represent stepping-stones.

While there are many examples of migrants' first jobs, or even of migration itself, being constituted as stepping-stones, or potential stepping-stones, these should not be overstated. The international experiences of many migrants are constituted of deprivation and hardship, with scant opportunities for knowledge acquisition, learning and enhancement of their curricula vitae. Instead, they may become entrapped, with their knowledge being undervalued, and few opportunities to add to this. Entrapment can occur in all three segments of what Portes and Bach (1985) term the 'triple labour market'—the primary, the secondary and the ethnic enclave. Employability prospects are likely to be different for migrants in each of the three sectors. The primary sector, with better-paid and more secure jobs, provides the best opportunities for formal training, simply because employers will have a greater incentive to invest in workers given their more stable workforces. It therefore offers the best means to enhance employability, even if it is initially more difficult to find employment in this section. Migrants may be more employable in the secondary labour market, not least because this is often strongly influenced by, while influencing, irregular migration (Piore, 1979).

Where migrants face obstacles to social or occupational mobility in the primary labour market, this is understood as the 'blocked mobility' thesis (Bonacich and Modell, 1980), but it also represents problems in their employability. Under such conditions, migrants are likely to be most employable in the enclave labour market, where there is also a strong presence of coethnic employers. However, this may also be a labour market segment which provides poor opportunities to learn from coworkers (particularly about the culture and language of the country of origin) and may become a form of entrapment.

A different perspective on labour market entrapment and stepping-stones is provided by the notions of ascription, acceptability and suitability (Jenkins, 2004: 153). Suitability emphasizes achieved or acquired characteristics; migrants have more power to change these by acquiring particular skills or knowledge of the national culture in the destination country. However, the social recognition of their suitability is mediated by ascription and acceptability. Ascription—who you are—is likely to influence perceptions of acceptability, or whether you fit into the networks and values of an organization. Migrants may be ascribed as outsiders, newcomers, or—in some cases—as ethnic minorities. Ascription applies to all migrants and not only to those who are conventionally considered to be unskilled. This is a process which can be observed in both the skilled and unskilled labour market segments, with Saxenian (2006), for example, reporting that many Asian scientists in Silicon Valley encountered glass ceilings in their employing organizations when they tried to move into senior management posts. Similarly within the health sector, all occupational grades are subject to ascription and acceptability barriers, even though we know more about the experiences of doctors and nurses (see Larsen et al., 2005; Raghuram and Kofman, 2002) than of cleaners and porters.

Migrants, like most newcomers, are also likely to be ascribed a peripheral position within work groups within organizations (Lave and Wenger, 1991). If they are to overcome ascription barriers and achieve acceptability, then this involves, as with all newcomers, moving 'incrementally along a continuum from the domain of stranger toward that of friend' (English-Lucek et al., 2002: 97). This may be critical to their employability and in shifting from potential labour market entrapment to labour market stepping-stones.

CONCLUSION

There are complex relationships between migration and employability. At an aggregate level, the structural and institutional features of the destination are important in determining the employability of migrants. Both the broad political economic differences between national states in relation to employment and welfare regimes and different approaches to migration policy are important in this. Within this broad context, there are also significant differences in the experiences of lead and trailing migrants and the outcome is highly ageist and gendered experiences of employability. The nature of migration itself also influences the employability outcomes. The most obvious differentiations are between labour migrants and refugees/asylum seekers and between registered and nonregistered migrants, both of which tend to be racialised. But the changing nature of migration, with more temporary and circular mobility, also has implications for employability. There is therefore a need to understand employability in terms of changing individual experiences within a broader canvas of economic and regulatory changes. Migrants are not passive dupes in this bigger picture but, to sharply varying degrees, individuals who can shape their employability, whether in terms of survival strategies or career building.

This chapter has also argued that employability has to be seen as being at the nexus of multilevel interests and actions. Key amongst these are the national state, firms and individuals. The interventions of national states are driven by both the competing economic needs of different sectors of the economy, and the discourses which surround these, and by the politics of migration. Moreover, the national state still strongly shapes the operating environment for firms, even in face of globalisation, as well as the conditions of entry, and the social and labour market integration of migrants and their employability. Firms, or more precisely industrial associations, are one of the main contributors to the national discourses about migration. They are heterogeneous, and the requirements of farming are likely to be very different from those of the health service or, say, biotechnology. However, there has been a relatively generalised shift to greater emphasis on flexibility, even if the extent and precise nature of this remains contested. Migrants can feature prominently in relation to firm strategies that seek to build competitiveness around different forms of flexibility. This has implications for the employability of migrants in terms of the types of jobs available to them and the propensity of employers to prefer migrant to nonmigrant workers. There are therefore wider employability implications for the labour force as a whole, as well as for migrants. Finally, while migration can be one way in which individuals respond to employability constraints in their country of origin, it can also be a positive strategy to add to their CVs and a valuable learning opportunity.

Outcomes do not always match aspirations and expectations. This is highlighted by the question of whether migration functions as a stepping-stone or a form of entrapment in the labour market. The answer is conditional, for there are differences in relation to age, gender, ethnicity and skills. These reflect the ways in which the notions of ascription, acceptability and suitability inform the ways in which firms, and fellow workers, approach the employment of migrant workers: their recruitment, allocation, promotion and colearning. Faced with barriers in the primary labour market, many migrants have sought jobs in the secondary or ethnic-enclave labour markets. These provide contrasting learning and employability outcomes.

Finally, this chapter has explored the usefulness of employability as a concept for understanding the relationship between migration and employment. First, and most obviously, it tells us about the labour market experiences of migrants—whether in quantitative or qualitative terms. Secondly, it opens up issues about social equity in terms of the experiences of migrants versus nonmigrants and amongst migrant groups. Thirdly, the employability outcomes of migration have to be understood not in terms of a single act of migration but of a possible cycle or cycles of migration. The most positive reading of this is that labour market entrapment in the destination country may turn out to be a stepping-stone to occupational mobility in the return country. However, a more negative reading may be that migration can lead to labour market entrapment in the destination country combined with little prospect of enhancing employability in the country of origin, especially when length of service or social connections are the drivers of career enhancement in the latter. Fourthly, the relationship between migration and employability has to be seen as multifaceted, involving not only the lead migrant but also trailing migrants, as well as the nonmigrant family members whose employability may be changed by the availability of remittances to invest in their education. In all these instances, we need to take a longitudinal perspective, to fully understand how the relationship between migration and employability is played out over the life course.

3 Methodological Challenges in Researching the Working Experiences of Refugees and Recent Migrants

Sonia McKay and Paula Snyder

This chapter reflects on some of the methodological challenges encountered over a three-year period in the conduct of two research projects on refugees and on recent migrants, exploring issues in relation to objectivity, trust and the 'hard to reach'. The first[1] was a thirty-two-month HE-ESF project exploring the barriers experienced by refugees in their routes into employment, to understand what the factors were that caused refugees to have poor labour market outcomes. As part of this research we also produced a short film, *All by myself*[2]. The second[3] was a twelve-month project, funded by the UK Health and Safety Executive (HSE), and its aims were to determine whether recent migrants to the UK were at greater risk in relation to their health and safety at work than were local workers. The chapter begins by noting that there is a wide body of literature looking at methodological issues related to research into the working lives and experiences of refugees and recent migrants, much of it emphasising the importance of using an appropriate methodology. However, it suggests that while the choice of method is important, it is crucial not to be trapped into one methodology but to remain open to different methods, as 'ultimately the choice of data collection modes will depend on the aims of the research, the research questions, the available resources, timescales and the characteristics of the survey population' (Bloch, 2007: 14). Thus, there is the need to maintain a flexible approach; to know as much as possible about the survey population prior to fieldwork; and to think carefully about the extent to which personal data are required. The latter is important not only because inquiry into the personal may be experienced as intrusive, but also because the collection of personal data from research subjects, whose legal status may be problematic, must be handled in a particularly sensitive and secure way. The chapter reflects on the context in which research is conducted, both in terms of the politicisation of migration and also in relation to the theoretical positions and experiences of the researchers themselves. Finally, the chapter looks at the issue of how to sustain trust relationships with research subjects, particularly in the conduct of longitudinal research and in the use of visual methods.

Both of the projects, on which we draw the evidence for this chapter, were primarily qualitative and were located in a direction of research that is socially engaged, whose aim is to understand why labour markets operate in a particular way and to connect with both the theoretical and policy issues arising from the employment experiences of refugees and recent migrants. Qualitative research was developed 'specifically to enable researchers to study social and cultural phenomena' (Myers, 1997) so that they might achieve 'the goal of understanding a phenomenon from the point of view of the participants and its particular social and institutional context', a perspective which 'is largely lost when textual data are quantified' (Kaplan and Maxwell, 1994). Indeed the 'growing appeal of qualitative interviewing has been made sense of, by considering it as a corrective response to the oppressive impact of scientism on our understanding of human experience' (Gregen, 2001). Within qualitative research a variety of methodologies can be explored, including comparative and longitudinal methods. But the methods need to be appropriate to the research subject. And whatever method is employed it is necessary to carry out in-depth exploratory work and extensive pretesting—especially in the ways in which irregular and insecure statuses may affect access to potential respondents (Bloch, 2004a: 3).

THE CONTEXTS SHAPING RESEARCH ON MIGRATION

In this chapter we have wanted not just to highlight some of the challenges we have encountered in the two projects but also to document how we responded to them. In doing so we first acknowledge a fundamental methodological challenge to capturing the essential contributions of research subjects, due to:

- The contexts which shape the working experiences of migrants and refugees; and
- The previously acquired expectations, cultural assumptions, form of language, and so forth, which the subjects bring to those experiences.

We believe that the resolution of this fundamental challenge requires the involvement of a breadth of experience among research teams and management boards; a sensitivity to the issues being investigated; a desire to engage in and to influence the debate; and a recognition that researchers do not need to be 'objective', in the sense of not holding any prior theoretical position at all, to be able to produce research which stands up to scrutiny, provided we articulate the often hidden values and characteristics of the researcher and their impact on the research process (Spalek, 2005: 408).

We identify three specific challenges in conducting research on refugees and migrants:

- Different degrees of objectivity and of engagement among researchers;
- Possible distortions of evidence arising from researcher/subject interface; and
- A continuum of ways of gathering data, from individual interview to statistically significant results.

Within any group of researchers individuals will hold differing positions that can influence the construction of key methodological tools, such as the interview schedule, the selection of interview subjects, the timing and place of interview and who conducts the interview. This makes it pertinent that the researcher articulates her or his position in approaching the research (Delphy, 1984; Gouldner, 1971), as:

'Research does not exist in a vacuum outside of the social world' and 'no researcher comes to a research project with a blank mind and no one is objective in this sense' so that 'it is not possible to divide researchers into those who are "insiders" and cannot remain objective and those who are "outsiders" and can produce valid, objective research'. (Temple and Moran, 2006: 9–13)

Thus researchers do not begin to formulate a hypothesis or construct a question while devoid of opinion or neutral as to the nature of the emerging data, but exactly for this reason a research team that is interdisciplinary or which is situated within a research environment that encompasses different disciplines may be better positioned to respond to the challenge of producing objective but engaged research. Here we have drawn on two complementary perspectives. One of us comes from a background as a researcher in employment law, adopting a critical perspective, assuming that 'social reality is historically constituted and that it is produced and re-produced by people' (Habermas, 1981). The other has a background in both research and in journalism and moved into film documentaries as a way of combining journalistic and research skills to document individual experiences. Based on these combined research experiences, we suggest that methods that are likely to lead to useful research outcomes need to borrow from different disciplines and to accept that there is no single perspective that is sufficiently broad, to provide all of the answers to the complex questions that arise, in trying to understand both the impact of migration on the individuals who move; on the societies which they leave; and on those to which they journey. For as Castles (2007: 353) notes, it is 'hard to do a useful study on any migratory phenomenon from a mono-disciplinary perspective'. For us, as for Castles, interdisciplinary 'does not mean putting them all together in a bland mixture, but rather building on and integrating the insights of the different approaches, to give a general understanding of migratory phenomena', what Brettell and Hollifield (2000) describe as 'talking across disciplines'. Thus, while coming from legal and journalism disciplines, to the study of migration and employment, we acknowledge that it is appropriate to explore other perspectives and other disciplines, to discover how law, economics, social relations and the study of demographics can be utilised to obtain deeper insights into migration and employment. Thus, while many of the issues explored in this chapter are more usually investigated by sociologists, the aim is not been to supplant those perspectives but to explore them through the filters of our own disciplines.

Research in the field of migration is inevitably highly politicised. The contested nature of the public and policy debates on migration requires of researchers at the very least that they challenge the 'unquestioned common sense' (Goldberg 1993: 41–43). As Castles comments, research on migration occurs in a context where 'fundamental ideas on the nature of migration and its consequences for society arise from specific historical experiences of population mobility and cultural diversity' while past experiences with internal ethnic minorities have helped shape current attitudes and approaches, so that this 'affects even the most critical researchers' (Castles (2007: 358). Government policy on migration and on asylum is also in a constant process of change, and in the UK this has led to a significant corpus of research that is government funded and policy driven and which, in the view of Castles (2007), 'may not only be bad social science' but 'it is also a poor guide to successful policy formation', particularly where it is designed to meet short-term political objectives. However, in contrast, the best such research can challenge existing assumptions and can draw open a curtain on how societies treat their most vulnerable members (Castles, 2007). Policy-led research also has the potential to facilitate access that otherwise might be difficult and provides social scientists with the resources to carry out empirical research on important emerging issues (Castles, 2007). Consequently, the solution is not to choose the one side or the other of the policy/academic divide, but to seek new ways of integrating them through the linking of empirical research to broader theories of social relations, structures and change:

> Thus researchers should relate not only to official policy-makers but also to non-governmental organisations and to civil-society groups of all kinds (both formal and informal). In the end, the aim must be to make it clear that historical and comparative principles, based on participatory methods as well as theoretical, historical and comparative principles, lead to more useful public knowledge than short-term policy-orientated studies. (Castles, 2007: 364)

The HSE research, which was aimed at developing a policy-led agenda on migrant health and safety, involved a wide range of nongovernmental organisations and civil-society groups, particularly through the medium of the advisory board (see following). We found that conducting policy-led research offered some advantages, particularly in relation to access. Key respondents in the HSE project, in particular employers and trade unions, were more willing to engage with the research than was the case in the refugee project, which was not policy driven. In the former, they could perceive an immediate advantage in their involvement in a developing agenda on the issue, whereas in the refugee project it was more difficult to convince such respondents of the direct value of the research to them. However, the same conclusions cannot necessarily be drawn in the case of worker participants, who may be more reluctant to take part in policy-led research, precisely because it has government support, in circumstances where their own experiences make them distrusting of public authorities. This is particularly the case in relation to non- or semicompliant workers (Tait, 2006).

In researching into migration and work we need to consider whether particular characteristics or experiences within the research team are essential to the conduct of the research. Feminist research advocates that there should be common characteristics between researcher and research subject, stressing 'the importance of establishing non-hierarchical relationships with interviewees' (Oakley 1981: 44) and this raises the question of who should conduct such research and under what conditions. Should researchers and the researched share commonalities, and if they should, how do we decide what these essential commonalities should be, in cases where issues of gender, class, race, nationality and age are all relevant? For these reasons, and reflecting on her research on black Muslim women post-9/11, Spalek suggests that a simple desire to empathise might encourage the researcher to focus on similarities between researcher and interviewee rather than on differences, even where it is precisely these areas of difference that are central to the key research questions (Spalek 2005: 413).

Mestheneos (2006) notes that 'overall the quality of the interview material generated tended to depend on the competencies and commitment of the individual interviewer rather than on their gender or (non-)refugee status', and indeed the fact that the refugee interviewer was often known to the interviewee influenced responses. Temple and Moran (2006) also caution against the automatic assumption that surface commonalities inevitably create nonhierarchical relationships and note the limitations of 'racial matching', as 'race is not the only relevant social characteristic'. For them the 'current trend towards employing one researcher from within a community, to represent community views, can be problematic from the outset, as 'community' is often narrowly defined in research (2006: 11; see also Jewkes and Murcott 1998).

Temple and Moran instead suggest that if the research is aiming to look at the needs of particular groups of people, they should have a say in how their needs are defined, while guarding against a tendency to assume 'that as "insiders" their knowledge is superior', as this 'can lead to a hierarchy of knowledge' which is not necessarily conducive to good research (Temple and Moran 2006: 9–11). Taking account of these views, it seems appropriate that while research teams might include individuals whose experiences are common to those of the research subjects, it is essential that the process does not consist simply of attempting to match researchers and researched.

While empathy is important, it should not risk being 'inclined towards producing flattering pictures of communities, which betray the substance of the temporary research friendships that are levered in the cause of research' (Mestheneos, 2006). And in any case confirmatory research relationships are not the only environment within which to conduct research. Knowles argues that antagonistic research relationships may be just as likely to produce valuable data (Knowles, 2006: 393–95).

ESTABLISHING AN ADVISORY BOARD

In both projects discussed here, as a first step, project advisory boards were established, consisting of individuals and organisations with particular expertise. They included representatives from refugee, migrant and community groups; from employers and employers' organisations; from trade unions; from groups working with refugees and recent migrants; and from fellow academics working in the field. The advisory boards were viewed as essential to the conduct of the research from initiation to conclusion, and indeed in both cases the first advisory board meeting took place within a month of the project's initiation, to ensure the active participation of the boards from the start of the research. For each of the projects the advisory boards represented a more formalised method of using nonprobability techniques, by allowing for multiple confirmatory networking. Advisory boards for each of the projects met at least four times a year. They were used to obtain comments and feedback on interview schedules, on the process of the filmmaking and on research reports as they emerged in the course of the study, and were also used to source potential interviewees or organisations or groups that might facilitate access to interview subjects. Throughout the research, the advisory boards provided robust but invaluable criticism. They were additionally useful in enabling the testing of emerging research findings with a knowledgeable and involved group of experts. Their different backgrounds also ensured that when sourcing interview subjects with their help, the range of gatekeepers was wide, guaranteeing that interviewees reflected differing experiences. Advisory board members were involved in the selection of the documentary filmmakers, and the filmmakers subsequently attended all of the meetings of the advisory board. This meant that the filmmakers were actively engaged within the research process, essential to ensuring that the documentary reflected the overall aims and objectives of the project, whilst at the same time accurately documenting three very different personal experiences.

CONSTRUCTING THE SAMPLE
AND CONDUCTING THE RESEARCH

The refugee research aimed to conduct interviews, with at least sixty refugees, three times over a two-year period. All interviewees had to already have the legal right to remain and work in the UK, having had their asylum claims accepted or having otherwise been given a right to remain. We specifically choose only to interview those with established refugee status, to exclude the possibility that poor labour market outcomes might be as a result of an insecure status. We were also to interview sixty black and minority ethnic (BME) workers, matched as closely as possible to the refugee interview group in terms of family circumstances, country of origin, age, education and area in which they resided in the UK. Both groups had to have their (or their family in the case of the BME group) country of origin in one of three areas: the Indian subcontinent, the Middle East or eastern Africa. In the UK they had to be residing in one of three locations: London, Slough or Birmingham. The aim of conducting interviews with refugees and with UK BME workers was to observe similarities and differences of experience and to investigate the strategies adopted within the BME group, in response to their experiences of labour market exclusion to see whether these were transferable to the refugee group. The production of the documentary film *All by Myself* was part of the project. The film was to be used to help disseminate the findings of the research, by focussing on the human stories of some of the people who had taken part. The film was expected to be used by groups like trade unions, employers or community groups working with refugees, but was to be produced in such a way that it could potentially reach a wider audience. 'Talking heads' were to be kept to a minimum, allowing the refugees' stories to dictate the film content.

In the second project, funded by the HSE, the essential requirement was that the sample of migrant workers should be sufficiently large to allow us to tentatively generalise from the findings. It was therefore agreed that we should interview at least 200 recent migrants who were working, or had worked, in one of six industrial sectors.[4] To ensure a reasonably wide geographical spread, the sample had to be reasonably equally spread between five regions of England and Wales[5] and had to reflect the size of the six industrial sectors. Interviews also had to be conducted with between 80 and 100 employers or labour providers, reflecting the same sectors and geographical spread of the worker interviews.

The primarily method of data collection in both projects was through face-to-face in-depth interviews. However, in the HSE research we also included an element of quantitative data collection, through questionnaires sent out to employers and trade unions and through a secondary analysis of the Labour Force Survey (LFS). We wanted some quantitative data, to provide a contextual frame for the qualitative data. In practice the data obtained through the questionnaires were limited, due to a relatively low response rate, while the LFS data, in relation to recent migrants and refugees in employment, were not large enough to provide statistically robust results. We thus utilised additional quantitative data, captured through a documentary analysis of more than 1,000 reports from HSE inspectors.

Thus, both projects used a mix of methods. The HSE project drew on qualitative and quantitative data, while the refugee project drew on qualitative and visual methods. Using mixed methods provided useful checks and balances in the research. However, we acknowledge that sometimes this is not possible, as research can be constrained by external demands and pressures, so that researchers' best efforts might be inhibited by organisational demands and how interview practices are located (Birch and Miller, 2002).

For the two projects the key questions in relation to the data gathering were:

- How to construct purposeful samples, in the absence of sufficient quantitative data to enable interviews on the basis of random samples; and
- How to engage with subjects who already saw themselves as 'over-researched'.

Bloch (2004a) and Tait (2006) have commented on how the lack of a sampling frame for refugees and recent migrants makes it difficult to construct a sample that would be statistically significant, and, in the conduct of the two research projects, we had to acknowledge that 'the impossibility of obtaining a random sample is irrelevant' (Harris, 2004: 17). For both projects we therefore constructed a sampling frame that took account of a variety of known factors, in relation to the composition of the refugee and migrant populations. For the refugee project the factors were age, gender, education, length of time in the UK, country/region of origin and location in the UK. Fixed numbers were set for each of the categories, and, where we had statistical data, the numbers reflected those proportions. In the HSE research the sampling frame was based on three factors: sector employed in, gender and region. Again, the sampling frame was constructed to reflect the known ratios. Interview subjects were identified through the use of gatekeepers and through snowballing. A possible weakness of the snowball method is that it has the potential only to include individuals within a specific network of people and to exclude others, leading to the possibility that research findings and subsequent analysis may therefore be subject to bias (Arber, 1993; Zulauf, 1999: 165). To avoid bias in selection it is important to use 'as wide a range of pathways as possible' (Bloch, 2004a). As indicated in the preceding section, although snowballing was used, we tried not to interview more than two individuals from the same workplace, and we aimed to engage with as wide a network of gatekeepers as possible, within the constraints of each of the projects, in relation to their duration and their budgets.

For the film we also wanted the people selected to be broadly representative of the profile of the project interviewees. Filmmaking production is usually scheduled into blocks of time allocated to preproduction, production and postproduction. However, *All by Myself* had a total of just seven and one-half filming days, spread over thirty-eight weeks, and this compounded the problems of trying to slot casting interviews, planning visits to locations and shoots in a sensible order. We also wanted to ensure that the second phase of filming would create a different feel from the first, to allow the viewer to appreciate that the film captured the stories of interviewees at differing times. This was assisted through the use of different settings and lighting techniques, although, in both phases of filming, decisions were taken to avoid dramatic angles and shots; and while extreme close-ups were used, extreme long shots were not, so as to avoid creating a sense of distance between viewer and interviewee.

The methods chosen thus had to respond to the specific aims of each of the projects. In the refugee project the aim was to understand whether the known poor labour market outcomes for refugees were as a consequence of time, location, ethnicity, skill and educational achievement, age, gender or other factors. By interviewing on a number of occasions the aim was to see whether labour market outcomes improved (or worsened) over time and whether the refugees' own assessments of their labour market situation transformed over time, and in what way.

For the film we wanted the people selected to be broadly representative of the profile of the project interviewees, and we wanted to conduct two phases of interviews, again to capture their stories organically rather than at a set point in time. The HSE project aimed principally to document the actual work experiences of interviewees at the time of the interview. However, it also aimed to explore whether, and to what extent, sector of employment, gender, age and status increased or reduced risks to workers' health and safety. In both projects the primary methodology was based on face-to-face, in-depth interviews, each generally lasting between sixty and ninety minutes. The aim was to understand the processes, and this is best be achieved by in-depth interviews (Arber, 1993; Zulauf, 1999: 162). We used a modified form of biographical interpretive method, with interviewees being asked to tell their biographical story from the moment of their arrival in the host country and then, in the course of the interview, being encouraged to reflect on their situation in their country of origin prior to migration/flight (Bertaux and Kohli, 1984; Breckner, 1998; Fischer-Rosenthal, 1995; Kohli, 1986; Rosenthal, 1993). Particularly in relation to the refugee research, we chose not to question individuals about their past in their country of origin, as we were aware of the risks of 'undertaking "trauma-exploration" or retraumatising participants in the zeal to exploit access to interviewees' (Chambon et al., 1998) and accepted that 'the past must be addressed as filtered though the present' (CCVT, 2000). But we also acknowledged that that by asking the same questions to each interviewee does not mean that the nature of the responses would be similar. We perceived that there were important differences in the interviews by gender and found, as did Mestheneos (2006:31), that men responded differently from women in how they told their personal stories.

From the outset of the refugee project, the research methodology had factored in the possibility of individuals dropping out from the sample over the course of the research. In order to militate against a significant attrition rate, the research team had endeavoured to keep in touch with interviewees through the methods detailed following. But we had always accepted that we would need to oversample in the first wave of interviews, if we were to have any chance of keeping around sixty interview subjects throughout the three waves of interview. At the beginning of the project it was difficult to predict what the dropout rate might be, and in retrospect it may be that we were overconfident about the numbers that could be tracked through the whole of the thirty-two-month period. We began by interviewing eighty-three refugees in wave one. We anticipated a dropout rate that would bring the numbers down to around seventy for the second wave and around sixty for the third wave. As it happened, in the second wave we interviewed sixty of the first wave. In the third wave the number dropped further to less than fifty. The reasons for this fall are complex and multilayered. In some cases it was just that we could not locate the individual. They had moved, had returned home or otherwise were not contactable. In one case, sadly, the individual had died. But we also sensed that there were other reasons why the number willing to be reinterviewed fell. In some cases individuals had 'moved on' with their lives. They had obtained jobs, commenced new training or otherwise were in a space when they no longer wanted to reflect on their past or to only be perceived as 'refugees'. And in those cases where they had obtained employment they were more constrained by time and either less able or less willing to give up their limited leisure time to researchers. Nevertheless, overall we would submit that the project was relatively successful in maintaining the cooperation of the refugee interviewees, compared generally with longitudinal research projects stretching over this period of time (Ruhs et al., 2006). Similarly, with regard to the film, when the first interviews took place, the crew did not know with certainty what interviewees would be doing when it came to the time to refilm and had just to hope that those chosen would continue to want to be involved. Happily they did and the second phase of filming was completed with interviews with all participants in the first phase.

THE CONCEPT OF 'HARD TO REACH'

Much of the literature on researching refugees and migrants conceptualises it as consisting of 'hard-to-reach' groups. However, often the reality is that it is not the 'them' that is 'hard to reach' but rather it is researchers, the 'us', who may incapable of stepping beyond our own constructed notions of self. Certainly if we consider the position of refugees, they represent a particularly overinvestigated and overtracked group. They will have been through a formal legal or administrative process to determine upon their claim; they have generally had close involvement with public bodies, be they welfare, health or education bodies; and they often are connected to established BME communities through country of origin, family or friendship ties.

Thus, there is no obvious structural reason as to why they are 'harder to reach' than any other group, unless the researcher constructs her/himself as essentially different and views the crossing from one reality to the other as problematic. Thus, as in any research project, there are difficulties to overcome and problems that need to be resolved, but these are not insurmountable. This is not to say that there do not exist objective reasons why the complexity of the working lives of refugees and recent migrants may mean that they do not see themselves as having the time, the energy or the desire to take part in research, unless they view the research objectives as relevant and as likely to encourage change. Mestheneos (2006) notes that many refugees have been 'overinterviewed' and therefore might be reluctant to participate in more research that does not appear to offer positive changes to their lives. It was also our experience that some refugee- and migrant-support organisations were less inclined to participate. This was not primarily because they might have concerns over the status of the individuals whose cooperation was being sought but it was because:

- They felt overresearched already;
- They had concluded, from past experience, that research did not lead to any positive outcomes for them or for the refugee/migrant communities in general;
- They were themselves very stretched, either in dealing with the multiple problems which refugees and recent migrant encounter, or in the case of support organisations, due to financial pressures they just did not have the capacity to deal with seemingly unending requests from researchers.

In researching the lives of individuals who have recently arrived we also had to think about the language in which the interviews should be conducted. In most cases decisions in reality are primarily pragmatic and take account of the amount of available funding, the timescale for the research and the methods to be used to analyse the findings. In the HSE project, we were unable to determine in advance which language groups would be included, as the sample was to be constructed on the basis of regional, gender and sectoral spread and not on the basis of nationality. All that we were able to do was to ensure that the main linguistic groups were covered.[6] We had also decided generally not to use interpreters, as we accepted that there is 'no one correct translation and that the translator is like Aladdin in the enchanted vaults: spoilt for choice (Bassnet (1994), and also that 'such people are often treated as conveyors of messages in an unproblematic way (Temple et al., 2006b: 39). We preferred, instead where possible, to employ fieldworkers, who could conduct the interview in the relevant language and who could then translate and transcribe. The advantages of this method are:

- It encourages the development of a more natural form of dialogue between interviewer and interviewee, which is not interrupted or constrained by the need to translate back and forth;
- It excludes one further stage in the interpretation of data and thus limits the potential for its distortion; and
- It takes account of the need for the interviewers to be trained, not just in the interview techniques but also to understand and engage actively with the research aims.

SUSTAINING TRUST RELATIONSHIPS

To meet the aims of the refugee project it was important that participants agreed to be interviewed on at least three occasions over a two-year period, and for those selected for the film, there was an additional requirement to participate in two phases of filming. This meant that we had a specific concern to develop long-term trust relationships with each of the interviewees, to encourage their continued participation. Other researchers have stressed the importance of building trust relationships, noting particularly that this 'takes time' (Hynes 2003), but, as Greenham and Moran (2006: 121–22) emphasise, it is easier to construct a 'high-trust culture' where communication is clear, where there is no hidden agenda and where there is a shared desire to achieve equality between all people. Trust includes guaranteeing confidentiality; it means having transparent systems in place that interviewees understand and approve of; and it means ensuring that identities are protected.

At the outset we acknowledge that the very action of conducting research influences the subjects of research and that therefore researchers do need to take account of 'which social considerations are required for a person to express him or herself in relation to another' (Knapik, 2006). As Cowles (1998) observes, interviews may encourage the interviewee to construct her or his story or to answer questions in a particular way, creating the need for 'sensitivity to the impact of research interviews on participants'. We are also aware that interviewees may interpret signals and actions and try to offer data they perceive are being sought. We identify two issues that researchers need to reflect on in relation to the potential for distortion of findings. These are: whether interview subjects should be paid and whether advice and support should be provided to interview subjects and what the nature of this should be.

In both research projects we paid migrant interviewees a small sum of money—around £20 per interview. Making payments in return for interviews may carry with it the risk that samples are skewed, if individuals press to be interviewed because they know they will be remunerated. However, researchers aware of this can take steps to ensure that the integrity of the sample is maintained. In the refugee project there were a few occasions (but remarkably few) where individuals would present themselves, asking to be interviewed, having been told to do so by friends who had themselves previously been interviewed. In these cases, if the individual's profile did not fit the sample, researchers simply would refuse to conduct the interview. But if it did fit, we were pleased to interview. There was no reason why the individual's story would be less honest than that of any other interviewee, simply because he or she was aware that a payment would be made. Lammers (2005), in her paper on the 'taboo of giving', makes the point that 'ultimately people decide what to tell, how to tell it, and what to hide or be quiet about' whether or not they have received a payment or other gift. And it is just as important to stress that the use of remuneration in ethnographic research is as likely to minimise bias (as to maximise it), by including people who would otherwise decline involvement, because they place a greater value on their time, energy and expertise (Thompson 1996), and furthermore that it provides some incentive to disadvantaged and overresearched groups (Patton, 2002; Kisson, 2006). Additionally, we believe that the payment of an incentive represents, on the part of the researchers, a clear recognition of the value of the interviewee's time to the project.

There is additionally a view that research evidence can be equally influenced by nonmonetary gestures, for example, through the relationships developed between researchers and researched, particularly in the conduct of longitudinal research (as in our refugee project). Maintaining contact with interviewees over several years inevitably results in researchers becoming, to some extent, involved in interviewees' lives and concerns. In both projects, interviewees on occasion did contact team members for advice or information. In situations where there are few or no support mechanisms, individuals will rely on any contacts they are able to establish, and it was inevitable that they should identify the researchers as a source of advice and information. This could include asking for references, checking job application forms, advising on ways of finding jobs, together with guidance on work-related problems and immigration issues. Lammers (2005) is supportive of this way of working, suggesting that trust is only generated where the researcher is prepared to 'enter into a personal relationship that involved sharing and giving'. Of course, researchers need to reflect on whether the development of more ongoing relationships could influence their research findings and as a minimum need to be aware of the fact that the continued involvement of some interviewees may be with the aim of maintaining these relationships. More importantly it also means that those who have less need for supportive relationships may be less likely to agree to maintain their participation in a longitudinal research project and thus their voice and stories will not be heard.

Maintaining contact during the whole of the research project and ensuring the continued collaboration of the interviewees also requires the institution of means of 'keeping in touch'. In the refugee project this was achieved in a number of ways. First, researchers were encouraged to keep in constant contact with interviewees. They sent letters/e-mails or texts to each interviewee immediately after the interview, thanking them for their participation. They then sent greetings cards with a short update to each interviewee a couple of months after each interview. This was to keep the interviewees informed of how the project was progressing and to assure them of the value to the project of their continued support. At the end of the first-year project, we hosted social events in each of the three geographical areas where the research was being undertaken to which all the interviewees within the area were invited. In the second year, contact was more regular since two waves of interview were organised, and, although we did not host another event, we did send an end-of-year letter and card to all of the interviewees, thanking them for their continued support and letting them know how the project had progressed. In the final year of the project, we also invited all those who had taken part in the research to come to the dissemination conference. Some did participate in the conference, with one participant also making a formal presentation at a plenary session. The film crew similarly had to maintain regular contact with interview subjects. A member of the research team had been present at all casting interviews, to help put interviewees at ease and to make sure that they understood that the filmmakers shared the objectives of the project, as the crew was aware that anyone talking about their experience of looking for work as a refugee in front of a camera was likely to feel exposed and vulnerable. The strong trust relationships that had already been built by the research team facilitated this process, and the working methods of the film crew strengthened this trust.

Within the projects, the fact that we did successfully develop solid trust relationships with many of the participants can be attributed to all of the factors set out above, but above all primarily to the quality and commitment of the researchers responsible for conducting the fieldwork. It is on them that the responsibility falls for sustaining trust relationships through the months or years of a research project, and it almost goes without saying that without socially committed researchers it is not possible to conduct socially committed research.

CONCLUSION

We hope the issues discussed in this chapter have been of interest to others working on projects related to refugees, migrants and work. Our aim has been to discuss some of the key challenges we have faced and to document the methods that we have used to respond to these challenges. At the same time it is important, as this chapter has argued, that research does record the stories truthfully and honestly. Engaging in research with groups whose economic or immigration status, or whose experience of racism and discrimination puts them in a position of particular vulnerability, imposes additional obligations on researchers as to how we tell, analyse, report and disseminate their life stories. What we do in our research can and does affect the way that such workers are viewed and in turn the way that host societies respond to their presence. Thus, the researcher engaging in such research has an obligation to reflect carefully on methodological issues and on the consequences of the use of different methods in constructing the research design and in conducting the research itself.

There is no template for how to do it, nor is there any point in time after which a researcher no longer needs to consider these issues. Whether beginning a research project for the first time or after many years of research, ways of working always need to be reflected upon; methods need to be reassessed; lessons need to be learnt from previous research; and above all, researchers have to accept the need continuously to learn from one another.

NOTES

1. McKay, S., Dhudwar, A., and El Zailaee, S. (2006). *Comparing the labour market experiences of refugees and ethnic minorities*. London: Working Lives Institute, London Metropolitan University.
2. *All by Myself* was produced by Glasshead Productions. Copies of the DVD are available from the Working Lives Research Institute. The film has been showcased at a number of events, including during Refugee Week 2006, at the Museum of London Refugee Film Festival and at various trade union events. The distribution of the film has been assisted by a grant from the Barry Amiel Trust.
3. McKay, S., Chopra, D., and Craw, M. (2006). *Health and safety and migrant workers in England and Wales*. London: HSE.
4. The sectors were: agriculture, cleaning construction, food processing, healthcare and hotels and catering.
5. The regions were: the East of England, London, the North East, the South West and South Wales.
6. Since the research was focusing on recently arrived migrants, the main linguistic groups were from Poland, Lithuania, China, Portugal and Latin America. There were also significant groups from sub-Saharan Africa, but often they already had a good knowledge of English and interviews could be conducted in it.

Part II

State Policies in Relation to Migrants and Refugees

4 Legal Frameworks Regulating the Employment of Refugees and Recent Migrants

Tessa Wright and Sonia McKay

This chapter examines the legal frameworks regulating the employment of refugees and recent migrants in the UK. Immigration and asylum laws primarily determine the extent of rights to enter and participate in the labour market, and these have become increasingly restrictive for some groups of migrants while encouraging others, as this chapter will show. The position of refugees and recent migrants in relation to antidiscrimination and employment laws and their understanding of their legal rights is also explored, drawing on recent research on refugees and the labour market as well as ongoing research on documented and undocumented migrants.[1] The chapter explores the short- and long-term implications of excluding asylum seekers and undocumented migrants from the labour market and the impact of successive and increasingly restrictive changes in immigration and asylum laws on the employment prospects of these groups.

HISTORICAL ACCOUNT OF IMMIGRATION POLICIES

Legislation to control immigration, particularly from the Commonwealth, was introduced in the UK in the 1960s. In the Immigration Act of 1971, and during the 1970s and 1980s, only strictly limited economic migration was permitted for skilled workers through a work permit scheme, and further requirements were imposed for those seeking family reunification. Restrictions on economic migration continued until the mid-1990s, when employers started putting pressure on the government to permit them to use more migrant labour. But at the same time the government was seeking to control the entry of asylum-seeking refugees, whose numbers had been increasing from the late 1980s. It maintained a distinction between those who arrived in the UK as asylum seekers and who were not allowed to work (despite their skill levels) and those who were eligible to work and seen as economic migrants (Baldaccini, 2003).

When the Labour government came to power in 1997, asylum and immigration procedures were considered to be failing, so they began a 'modernisation' of immigration policy, concentrating on improving administrative procedures and deterring asylum seekers, with the Asylum and Immigration Act 1999 (Flynn, 2005). The government's *Secure Borders, Safe Haven* white paper in 2002 (Home Office, 2002) went further in developing its policies on 'managed migration', referring to the economic and social benefits of migration, but also the need for the social integration of migrants, community cohesion and routes to gaining citizenship of the UK. Flynn (2005) argues, however, that migration policy was being shaped by the needs of business both for skilled labour and unskilled labour, and the rights of migrants in relation to length of stay, family reunification and so on varied according to the scheme under which they entered and depending on employer demand for labour.

THE LEGAL FRAMEWORK FOR REFUGEES, ASYLUM SEEKERS AND ECONOMIC MIGRANTS

The main routes for immigration into the UK are as an asylum seeker, for family reunion or for work. The 1951 Geneva Convention relating to the Status of Refugees,[2] as amended by the 1967 Protocol, guides the obligations of the UK in relation to applications for asylum. While these instruments set out a basic definition of a refugee, and the standards of economic and social rights to be accorded to refugees, there is no internationally agreed set of procedures to determine which persons will be granted refugee status. Neither the Refugee Convention nor the Refugee Protocol defines key terms, such as 'owing to a well-founded fear', 'persecution', 'membership of a particular social group' or 'political opinion', leaving the courts and tribunals in individual signatory countries to interpret these terms differently, in accordance with national and supranational interests. Since the Amsterdam Treaty 1999, policy in the UK has been influenced by the movement towards the harmonisation of asylum policy and practices within the European Union and the development of a common European asylum system.

While the Refugee Convention remains the cornerstone and the only international instrument for the protection of refugees, the UK has developed a body of asylum law with at least six major legislative changes to the asylum system over the past decade. Under successive governments, asylum legislation has been justified within a 'threat/defence' framework, assuming a direct correlation between numbers of asylum applicants and economic 'pull' factors such as welfare benefits and employment, which is presented as a threat to the welfare of nationals, who are often portrayed as victims at risk of being displaced from the labour market and social security, health care and public housing provision by asylum seekers and migrants. The objective of asylum policy therefore has been to deter applicants with economic motives through a number of measures.

As part of the clampdown on perceived abuses, access to public services and the labour market has become progressively contingent on an individual's immigration status and habitual residence tests. Therefore, while access to services and employment/training opportunities for those seeking asylum is highly restricted, those granted refugee status are able to claim welfare benefits, access public services and take jobs on the same terms as British nationals.

In 2005 the UK government published its five-year strategy for asylum and immigration (Home Office, 2005), which continued its focus on tightening up the asylum system to deter those it perceived as 'economic migrants' rather than 'genuine refugees'. Asylum seekers are denied access to the legal labour market, and the government's position is that the integration of asylum seekers can only begin once they become refugees, although there is a scheme for asylum seekers to undertake volunteer community work (Ensor and Shah, 2005). Once asylum seekers are granted refugee status they can work legally. In the past, refugees were granted indefinite leave to remain, but since august 2005 refugees and people with humanitarian protection (those who do not qualify as a refugee but who, if returned to their country of origin, would face a serious risk to life or person such as the death penalty, unlawful killing or torture, inhuman or degrading treatment or punishment) are granted five years' leave which is renewable. Permanent settlement is then only granted to those refugees who, after five years, are still eligible to remain in the UK.

Currently there are provisions under the immigration rules for spouses, fiancé(e)s, unmarried and same sex partners of those who are settled in the UK, or in a category leading to settlement, to come to the UK and settle. Dependent children, aged under 18 and unmarried, can also apply to join their parent(s) and settle. There are very limited opportunities for other categories of relatives, so grandparents, for example, must be wholly or mainly financially dependent and have no other relatives in their country who could support them (Baldaccini, 2003). At the time of writing some categories of work-permit holders could be joined by their spouse and dependent children, but these rules are due to change with the introduction of the new Points-Based System (see following). In addition, the government is proposing that in future family reunion will be predicated on the new family member (for example a prospective spouse) having a sufficient degree of fluency in the English language prior to his or her migration.

The work-permits system has represented the main mechanism for managing labour migration into the UK. In recent years it expanded considerably, with several new schemes introduced, but starting in 2008 all existing visa and work-permit systems will be replaced with a single points-based system, made up of five tiers. Tier 1 (introduced in the first quarter of 2008) includes only highly skilled professionals and entrepreneurs and offers a route to settlement, possibly after two years (Home Office, 2006a). Tier 2 is for skilled workers with a job offer from a UK employer, who may apply for settlement after five years' residence, and was to be introduced from the third quarter of 2008. Tier 3 is for low-skilled workers, limited by quota, where UK and EU labour is not available and offers only temporary residence for a maximum of twelve months, with no right to bring dependants or switch to another route. Tier 4 is for students, with leave tied to a sponsoring educational institution and limited to the duration of the course, but most will be entitled to bring their dependants and work part time. Tier 4 is intended to start at the beginning of 2009. Tier 5 (to begin in the third quarter 2008) covers youth mobility schemes and temporary workers who might not qualify under Tier 2 but who are allowed into the UK for cultural, charitable, religious or international development reasons (Home Office, 2006a). These proposals have been criticised for potentially exacerbating the vulnerability of migrant workers and their families, as critics fear that the effect might be to curtail legitimate labour migration routes and increase the likelihood, in the words of the chief executive of the Immigration and Advisory Service[3], that 'more workers will be sucked into the economy via smugglers and traffickers, with appalling consequences of exploitation' (United Nations Association–UK (UNA-UK), 2005).

Until the introduction of the five-tier scheme in 2008, the work-permit scheme allowed employers to recruit people from outside the EEA. In addition, there were specific schemes for certain types of worker. The Highly Skilled Migrant Programme (HSMP) allowed foreign workers outside the EEA to enter the UK in order to seek work without already having a job with a UK employer, as long as they had gained sufficient points based on criteria such as qualifications, previous earnings, age and experience of studying or working in the UK, and from December 2006 there was an additional English-language requirement. The Sectors-Based Scheme (SBS) was introduced in May 2003 to address shortages in lower skilled occupations, in particular food processing and hospitality. In July 2005 the hospitality sector was withdrawn from the scheme, on the basis that the required labour could be drawn from the enlarged European Union (Home Office, 2005), and the scheme for the food manufacturing sector was extended on 1 January 2007 for Bulgarian and Romanian nationals only.[4] The third main scheme was the Seasonal Agricultural Workers' Scheme (SAWS), which allowed low-skilled workers from outside the European Economic Area (EEA) to come to the UK for seasonal agricultural work, with an overall annual quota of 16,250 for 2007. However, the government intended, from 1 January 2008, to recruit only Romanian and Bulgarian nationals under SAWS. In 2007 a minimum of 40 per cent of the annual quota was allocated to Romanian and Bulgarian nationals, with the remaining 60 per cent for students from non-EEA countries.[5] The last of the schemes was the Working Holiday Makers' Scheme, which allowed Commonwealth citizens aged between 17 and 30 to come to the UK for an extended holiday of up to two years, during which they could work, but only if it was incidental to the holiday. The source countries were predominantly the 'Old Commonwealth', with Australians and South Africans accounting for two-thirds in 2005, although the numbers from Ghana, India and Malaysia have risen substantially (Salt, 2006: 86).

Domestic workers have also been allowed to come with their employer to the UK, and were normally given permission to stay for six to twelve months, depending on the employers' length of stay. They had the right to change employer to another job as a domestic worker in a private household, but had to notify the Home Office. They could apply for an extension to their stay, and after five years' continuous employment could apply to stay in the UK indefinitely. However, these rights are now threatened by the proposals for a points-based system, which will allow migrant domestic workers to be brought into the country by employing families as 'domestic assistants' for six months only. This would not be renewable and the workers would be expected to leave the country at the end of this period. The proposals are being challenged by Kalayaan, the migrant domestic worker campaign group, which points out that evidence of high levels of abuse and exploitation, of mainly female domestic workers, had led to the introduction of the current rules, following a regularisation programme for domestic workers in 1998 and 1999. This gave domestic workers the opportunity to regularise their status and therefore change employers. However, these gains would be jeopardised by the proposed changes.

In addition to the previous categories, the number of students coming to the UK has risen dramatically from just fewer than 60,000 in 1994 (Office of National Statistics (ONS), 2004) to 330,060 non-UK students in HE in 2005–06 (Higher Education Statistics Agency data).[6] Students have the right to work for a limited number of hours, and many take the opportunity to work, in some cases in excess of the permitted hours.

The UK government's policy on migration shifted once it had a measure of the extent of migration from the new member states (principally the A8 countries) into the UK. On 1 May 2004, when the A8 countries of Central and Eastern Europe[7] joined the European Union, nationals of these countries were granted the right to work in the UK, although they were required to register under the Worker Registration Scheme. By the end of 2007 more than 700,000 workers had registered under the scheme. These recent migrants are seen as 'solving' the problems presented by demographic changes and a growing ageing population. But racist discourses have increasingly informed the debate on migration, fuelling the idea that traditional migration routes, mainly from the developing south, could be replaced by migrants coming from Central and Eastern Europe whose longer term integration may be conceived as less 'problematic' or who are expected to be only short-term migrants.

At the same time as reforming the routes to labour migration, the government is taking an increasingly 'tough' approach to border controls and 'illegal' migration, including measures such as the introduction of compulsory identity cards, containing biometric data, for non-EEA nationals living in the UK (Home Office, 2006b). This tough approach has been driven by a desire to enhance public confidence in the immigration system and to address public concern about perceived high numbers of immigrants, fuelled by sections of the popular press (Ensor and Shah, 2005). The most recent proposals further stress the distinction between 'good' (perceived as beneficial to the UK economy) and 'bad' (said to be abusing the system) migrants, who are equated with those without permission to reside or work. The stated aim is 'to widen the gap between the experience of legal and illegal migrants' (Home Office, 2007: 5) by making it easier for legal migrants to come to the UK, and for some to settle, but further denying access to benefits and services for 'illegal' migrants. The Home Office is proposing a series of sanctions for 'illegal migrants' appropriate to the level of harm that it perceives is caused; so, for example, those involved in criminal activity would face prosecution and removal, whereas those who have stayed beyond their visa or permit would face lesser penalties (Home Office, 2007). Specific measures include a 'watch list' of individuals to be provided to government departments and lists of overstayers for checking with employers (ibid.: 22). The Home Office is also stepping up enforcement action in relation to employers of undocumented migrants and increasing the penalties on employers who are caught (see following, Access to the Labour Market).

There has been much criticism of the use of the term 'illegal' to describe migrants without full authorisation to reside or work, with many believing that most undocumented migrants are not criminals and that defining people as 'illegal' can be regarded as denying them their humanity and, furthermore, labelling 'illegal' asylum seekers who find themselves in an irregular situation may further jeopardize their asylum claims (Platform for International Co-operation on Undocumented Migrants (Europe) (PICUM), 2006). A more useful categorisation that takes account of the complexity of immigration status and permission to work are the terms 'compliant', 'semi-compliant' and 'non-compliant' developed by Ruhs and Anderson (2006), where compliant migrants are legally resident and working in full compliance with the conditions of their immigration status, noncompliant migrants are those without the right to reside in the host country and semicompliance indicates a situation where a migrant is legally resident but working in violation of some or all of the conditions attached to his or her immigration status.

Government policy is moving further away from the reforms advocated by organisations such as the Joint Council for the Welfare of Immigrants (JCWI), the Institute of Employment Rights (IER) and the Institute for Public Policy Research (IPPR) that call for according explicit rights to migrant workers and emphasise their human rights. There are also increasing calls for a regularisation programme for irregular migrants, with a number of trade unions, together with the TUC, calling on the government to regularise the position of undocumented migrant workers.

Although the UK government is opposed to one-off amnesties for undocumented workers (JCWI, 2006), there is already a permanent system of regularisation for those who have been in the country continuously for 14 years, regardless of legal status, and for families with small children who have been in the country for seven years (Levinson, 2005). These applicants are given indefinite leave to remain, with exceptions made only where there are serious concerns, such as a criminal history. Between 1996 and 2005 some 33,680 long-residence concessions were granted, an average of around 3,300 a year, with a high point of 9,205 in 2003 and a low point of 1,400 in 1999 (House of Commons, response to questions, 17 July 2007).

When the UK granted free movement of workers to nationals of the A8 central and eastern European countries in May 2004, many workers were in effect regularised. Analysis of Worker Registration Scheme data shows that by December 2004, 26 per cent of applicants had been in the UK prior to accession and a further 12 per cent did not give their date of arrival (Portes and French, 2005). While some will have been in the UK legally as visitors or students or working legally with a work permit, others are likely to have been working illegally (ibid.).

In April 2007 the Border and Immigration Agency took over responsibility for immigration control, work permits, nationality and asylum from the Immigration and Nationality Directorate but remains part of the Home Office. Both indigenous and migrant workers, including refugees, require a national insurance number before they start work, and these are allocated by the UK government Department for Work and Pensions.

ACCESS TO WELFARE

The Immigration and Asylum Appeals Act 1993,[8] the Asylum and Immigration Act 1996 and the Housing Act 1996 were the first of a series of measures designed to erode the rights of asylum seekers, and other non-settled migrants, and had the cumulative effect of restricting access to local authority housing and withdrawing entitlement to social security benefits, by linking eligibility with immigration status. Specifically, the 1996 act demarcated asylum seekers into two groups of applicants, restricting state support to those who claimed asylum at the port of entry.[9] The immediate 'knock-on' effect was to leave many asylum seekers homeless and destitute, forced to live a hand-to-mouth existence and susceptible to seeking work in the informal economy. However, following a legal challenge, the regulations were held to be *ultra vires* and responsibility for supporting destitute asylum seekers shifted from central to local government under the provisions of the Children's Act 1989 and by virtue of the National Assistance Act 1948, stating that local authorities had a legal obligation to provide financial support and accommodation to asylum seekers as people 'at risk' and 'in need of care and attention' (*R v Westminster Council and Ors* [1997]). These localised arrangements were eventually replaced by centralised support arrangements, the National Asylum Support Service (NASS), which assumed responsibility for providing accommodation and support to destitute asylum seekers.

The centralisation of support provision was the core theme of the Immigration and Asylum Act 1999 (IAA) and the Nationality, Immigration and

Asylum Act 2002 (NIA). The IAA 1999 and NIA 2002 made sweeping, contentious changes to the way asylum seekers were supported through a noncash centralised system based on vouchers and geographic dispersal. The voucher system, introduced under the 1999 act, was socially divisive and created unnecessary hardship by forcing asylum seekers to live on 70 per cent of the basic levels of income support (The Asylum Coalition, 2002). A Home Office report on asylum seekers' experiences of using vouchers pointed to difficulties in travelling to collection points as well as designated shops, the inadequacy of vouchers to pay for certain items and the inability to save (Home Office, 2002). The voucher system was abolished in April 2002.

In order to relieve housing pressure in London and the South East, asylum seekers have been dispersed to locations around the country on a no-choice basis and often to areas with no support infrastructure. The dispersal policy has been criticised as a 'massive experiment in social engineering' (Refugee Council, 2000) which fails to consider asylum seekers' needs, for example, access to essential services such as legal representation and ethnic minority/community networks. The effect of the dispersal policy, and the establishment of a network of accommodation centres, has been to resettle individuals in areas with a limited history of immigration and quite often within hostile areas, creating marginalized and dependent refugee communities (Zetter and Pearl, 1999) to the detriment of race relations and integration.[10] Racist incidents against asylum seekers appear to be on the rise. In October 2001, 112 incidents of racial harassment were reported to the National Asylum Support Service by asylum seekers (NASS, 2001), and local police figures from Hull reveal that on average one racially motivated incident is reported every day (Regan, 2000).[11]

Although the dispersal policy is not framed in punitive terms, asylum seekers who opt to live with friends and relatives in London or the South East, rather than be dispersed elsewhere, forfeit their right to accommodation and receive a cash-only support package. The lack of choice on regional settlement and the withdrawal of accommodation (where applicable) raise serious concerns of overcrowding where individuals choose to live with friends, and the lack of permission to work could increase the risk of homelessness and illegal working where private landlord accommodation is charged at exorbitant prices. In relation to employment opportunities, the implications of the dispersal policy in moving asylum seekers away from dense cities and regions are to remove them from the informal networks that are important in terms of future job seeking and employment within refugee communities. Interestingly, those who were able to choose their location were more successful in finding work than those who were dispersed without choice (Bloch, 2004a).[12]

The Asylum and Immigration (Treatment of Claimants) Act 2004 (AITCA) restricted asylum seekers' access to asylum appeals, created new penalties for people arriving to the UK without documentation, created a new criminal offence of failing to cooperate with detention and removal and increased the powers available to remove asylum seekers on 'safe third country' grounds. The AITCA also rather controversially withdrew support from families at the end of the asylum process. A failed asylum seeker with family (who refuses without 'reasonable excuse' to return to his or her home country voluntarily) is now the fifth class of person ineligible for support or assistance from local authorities. The Immigration, Asylum and Nationality Act 2005 grants social housing to persons subject to immigration control and makes provision for refugee integration loans to help refugees with the costs of integration.

In relation to migrant workers, the scheme for A8 nationals gave them no right to income-related state benefits until they had acquired twelve months 'habitual residence' requiring them to be in work continuously for at least twelve months.

ACCESS TO THE LABOUR MARKET

Although much of 'third way' politics espoused by the Labour government since 1997 focuses on tackling the mechanisms of social exclusion, some groups are actively excluded from taking advantage of 'social goods' such as employment because of institutional barriers which cause and reproduce the very patterns of inequality and disadvantage that are the core concerns of the discourse of 'social inclusion'. Until July 2002, asylum applicants were able to apply for work permission six months from the date of their application. That concession was withdrawn on the basis that employment acts as a 'pull factor' and that the majority of asylum decisions are made within six months. Since then, only individuals who obtain refugee status have the same employment rights as all UK citizens. Asylum applicants who were allowed to work or had applied for work permission prior to July 2002 continued to have the right to work. However, this is limited to the main applicant; dependants of the applicant are not given permission to work until refugee status or exceptional leave to remain (ELR) has been granted. These initial periods outside the labour market may have longer term adverse effects on the labour market participation of refugees (Refugee Council, 2005; Valtonen, 1999). Recent unemployment figures indicate that refugees are comparatively disadvantaged, but it is not clear how far that can be attributed to the change in policy.

The Asylum and Immigration Act 1996 made it an offence for employers to knowingly employ someone without, or with false, documentation. In 2004, in order to clarify the system, new guidance was issued on the types and combinations of documents employers were required to check under Section 8, as proof of a new employee's entitlement to work,[13] and these new arrangements came into force on 1 May 2004. But this did not mark the end for legislation aimed at employers of migrant labour. The

Immigration, Asylum and Nationality Act 2005[14] created a new civil penalty for the offence of employing a person subject to immigration-control restrictions. An employer found to be employing undocumented workers faced a fine of up to £5,000 per worker, and from March 2008 the penalty doubles and the requirements on employers regarding the checking of status documentation increase. While the legislation includes a statutory defence for employers to negate charges of employing undocumented workers, the duty to inspect papers has already caused concerns and difficulties. A survey of ten employers reported reluctance among employers to recruit asylum seekers or refugees because of the difficulties in verifying the accuracy of documentation, given the improving quality of forgeries. The time and costs involved in undertaking such checks were also highlighted as a deterrent, when compared to the alternative of recruiting migrant workers with work permits (Employability Forum, 2004). Among the recommendations put forward by employers was a 'fast-track' system and national ID cards for non-UK nationals as alternative methods of verifying the immigration status of individuals applying for work. However, the proposal to introduce ID cards has received criticism for its consequences for race relations (Liberty, 2004;[15] CRE, 2004,[16] JCWI[17]). As the JCWI stated, the negative interaction between employment issues and immigration policy has been observed since the implementation of Section 8 of the Asylum and Immigration Act 1996, and new employment offences will only serve to reinforce discrimination against British ethnic minorities in the labour market, particularly as the 2005 act extends employers' obligations from the mandatory check at the point of hiring, to monitoring that status throughout the entire period of employment (JCWI, 2005).

EMPLOYMENT LAW FRAMEWORK
FOR MIGRANTS AND REFUGEES

In the UK the national employment law framework is based on the existence of a valid and legally enforceable employment contract, and this contract remains the primary determinant of entitlement to statutory employment rights. The fact that employment rights are dependent on employment status and on the type of contract under which employment occurs means that some workers are excluded from basic rights such as protection over matters like dismissal, redundancy, maternity, paternity and parental rights. Migrants working with authority have access to the same employment rights as UK workers. This is also the case for refugees whose asylum request has been approved. However, for migrants who are noncompliant or semicompliant (working without or beyond their authorisation) or for asylum seekers awaiting a decision on their case or asylum seekers whose case has been rejected, under UK law their employment contracts are 'illegal' and non-enforceable. McKay et al., (2006) found that those working under illegal contracts were most likely to be paid less than the national minimum. Furthermore, whether or not a worker can enforce employment rights is dependent on the level of support and information available to the worker and is premised on knowledge of rights. Although this applies generally to all workers, given their labour market vulnerability, unauthorised migrants are disproportionately affected by the general failure of UK employment law to guarantee basic employment rights to all workers. Thus, if workers are employed, *to any extent*, in circumstances where they have no legal authorisation to work, they place themselves in an illegal contractual situation, the effect of which is to deny to them the right to enforce any statutory or contractual employment rights. This outcome is not directly a consequence of specific legislation but is based on interpretations of contract law made by the courts and which set precedent. A recent court ruling in this area was in the case of *Vakante v Addey and Stanhope School [2005] ICR 231*. Here, the UK Court of Appeal held that a Croatian national, who had been working in breach of immigration rules which did not authorise his right to work, could not pursue, against his employer, a claim of discrimination under the Race Relations Act 1976. In making its ruling the court endorsed the approach to the operation of the doctrine of illegality set out in the earlier case of *Hall v Woolston Hall Leisure [2001] ICR 99*. More recently, in the case of *Klusova v London Borough of Hounslow [2007] EWCA Civ 1127*, the Court of Appeal has held that dismissing an employee due to a mistaken belief that the individual no longer had a right to work in the UK did not of itself amount to an unfair dismissal. The effect of these rulings is that a non- or semicompliant migrant worker is unable to enforce any employment rights that are dependent on the making of an individual claim.

As a result of research and high-profile legal cases into recruitment processes and workplace relations, there has been a growing awareness of discrimination and racism in society more generally and of the consequent need to break down barriers and change attitudes through legal and nonlegal measures, such as antidiscrimination laws, voluntary policies, public awareness campaigns, positive action and collective worker action. Although legislation has expanded protection and benefits for workers, this must be seen against a background of shrinking public assistance for victims seeking access to justice. Various pieces of legislation exist to protect workers from discrimination and unequal treatment on the basis of race, nationality, gender, disability, religion, sexual orientation and age, in addition to laws protecting rights such as the national minimum wage, health and safety, working hours, trade union membership, maternity, paternity and paid leave, and so on.

However, legal measures form only part of the strategy to combat discrimination in employment. Ensuring individuals are aware of their rights and have the appropriate support and resources to lodge complaints also needs attention. A Department for Trade and Industry (DTI) report found that 91 per cent of respondents, in a study examining awareness and knowledge of antidiscrimination legislation, when prompted, were aware of the right to be treated fairly regardless of race, gender and disability. Unprompted and prompted awareness of antidiscrimination legislation are highest amongst white-collar (especially managerial and professional) occupations, in the public administration, health and education sectors, in larger workplaces and among trade-union members. Research demonstrates that refugees and recent migrants are concentrated in sectors characterised by low pay, poor working conditions and casual/temporary employment (Bloch, 2002) and thus are among the most disadvantaged and *perhaps* the most unlikely to challenge racism and inequality in the workplace.

To respond to the charge that the threat of penalties against employers who take on unauthorised workers might encourage discrimination, guidance by the Home Office in 1996, together with a Code of Practice under Section 22 of the Immigration and Asylum Act, was issued in 1999 to remind employers of their obligations not to discriminate in recruitment or employment under the Race Relations Act 1976. Even so, previous studies examining the labour market barriers faced by refugees have consistently highlighted the prevalence of discrimination.

According to a joint paper by the Commission for Racial Equality (CRE), JCWI and the Refugee Council, these 'internal immigration controls', carried out at the point of access to benefits, services and employment, have created a culture of suspicion which has increased the potential for unequal treatment for some people, in particular those from ethnic minority, immigrant and refugee communities (CRE, 1996). According to the CRE research, some employers have misunderstood and wrongly applied s.8 of the act. In the examples given, existing employees have been dismissed and job applicants have been discriminated against because employers have misread passport stamps or only accepted certain documents, such as passports, as proof of the right to work.

CONCLUSION

This chapter has shown that an increasingly restrictive legal regime for refugees and economic migrants in the UK has consequences for their labour market rights. It has shown that tight rules on admission and on access to work make workers vulnerable and less able to exercise their employment rights, even where they exist. But it has also suggested that the legal system itself operates to encourage this situation and to deprive refugees and migrants of their rights to challenge employer abuse. Additionally, the imposition of an obligation on employers to check on individuals' immigration status—described as a form of 'internal immigration control'—can have a discriminatory impact on the employment of all migrants and ethnic-minority workers when employers favour, for employment, those who they believe will require fewer checks.

The system of migration in the UK could be viewed as being at a crucial stage. The extent to which migration and asylum become increasingly regulated and restricted could turn back decades of struggle for the integration of new arrivals to the UK. The diversity of the UK's labour force and the experiences and understandings that migration can bring into the labour market will be challenged by legislation which no longer sees migrants as potential permanent residents but as 'guest' workers whose presence is 'tolerated' only so long as they are seen as plugging holes in a system which has failed to up-skill and qualify sufficient numbers of workers to fill the available jobs. In such circumstances it is difficult to advance an 'upbeat' conclusion, when future migration policy increasingly appears to be based on the offer of short-term residency rights, perhaps accompanied by large-scale deportations of those without the right to remain.

NOTES

1. Further details of the seven-country Undocumented Worker Transitions project can be found on the Website at www.undocumentedmigrants.eu.
2. Available at: http://www.unhchr.ch/html/menu3/b/o_c_ref.htm (last accessed 19 January 2005).
3. A UK charity providing representation and advice in immigration and asylum law.
4. IND Web site 23 March 2007.
5. Immigration and Nationality Directorate Web site 23 March 2007, www. workingintheuk.gov.uk/working_in_the_uk/EN/homepage/work_permits0/ seasonal_agricultural/general_information.html.
6. UKCISA Web site, http://www.ukcosa.org.uk/about/statistics_he.php, accessed 8 January 2008.
7. Czech Republic, Estonia, Hungary, Latvia, Lithuania, Poland, Slovakia and Slovenia.
8. http://www.legislation.hmso.gov.uk/acts/acts1993/Ukpga_19930023_en_1.htm.
9. In particular the act sought to reduce the number of asylum applications by restricting support to in-country applicants. Thus, provision was made to exclude in-country and failed asylum applicants, including those who were pending an appeal, from claiming means-tested welfare benefits.
10. See 'Asylum Seekers Unite to Fight Racial Abuse', *The Guardian*, 4 January 2005, http://www.guardian.co.uk/print/o,3858,5095279-110414,00.html (last accessed 20 January 2005).
11. Regan, S. (2000). 'Everyday Lives of City's Refugees', *Hull Daily Mail*, 9 November.
12. Available online: http://www.ippr.org.uk/research/files/team19/project183/ippr WPRefEmp.PDF (last accessed 16 February 2005).
13. Home Office (2004). Changes to the law on preventing illegal working: short guidance for UK employers, IND Corporate Communications: London.
14. The act also reduces the range of immigration decisions which attract a right of appeal; restricts appeal rights of visitors (family members), increases the information which may be required of passengers to the UK; requires passengers to provide biometric data; plans to grant social housing to persons subject to immigration control; introduces integration loans.

15. Available online: http://www.liberty-human-rights.org.uk/privacy/id-cards-2nd
-reading-commons.pdf (last accessed 16 February 2005).
16. Available online: http://www.cre.gov.uk/pdfs/id_cards.pdf (last accessed 16
February 2005).
17. Available online: http://www.jcwi.org.uk/campaigns/IDcards/lordsbriefing.html
(last accessed 16 February 2005).

5 The 'Visibly Different' Refugees in the Australian Labour Market

Settlement Policies and Employment Realities

Val Colic-Peisker

This chapter explores the intersection of labour market experience of permanently protected humanitarian entrants who arrived in Australia in the 1990s–2000s and the Australian immigration and settlement policies pertinent to their experiences. The data used come from a three-year research project focused on employment outcomes of three refugee groups (ex-Yugoslavs, black Africans and 'Middle Easterners') in relation to their racial and cultural visibility and therefore a potential for being discriminated against in the labour market. One hundred fifty respondents (fifty from each group) had high human capital profile, but the employment outcomes, after on average five to seven years of residence, were poor, which is consistent with government statistics that show humanitarian entrants having considerably poorer employment outcomes than other immigration streams. This is usually blamed on low human capital and refugee trauma, but this project indicates that other factors may be at work. Among them, labour market segmentation, which allocates visibly different refugees into undesirable jobs, may be the main one. Labour market segmentation results from systemic discrimination (e.g. nonrecognition of qualifications) and is reinforced by mainstream prejudices and negative stereotyping of racially and culturally different immigrants and refugees by employers.

INTRODUCTION

Australia is one of only ten Western countries that have planned annual refugee quotas. This 'offshore'[1] humanitarian immigration program admits 12–13,000 people a year on average, which is usually under 10 per cent of the Australian total annual immigration, dominated by the skilled category (Commonwealth of Australia, 2007). Although the refugee[2] intake is ostensibly driven by a need to protect some of the most vulnerable people, the criteria that drive the skilled intake creep into the humanitarian intake (Refugee Resettlement Working Group (RRWG), 1994). In other words, better educated, young and healthy refugees with at least some knowledge of English—and therefore with a higher 'resettlement potential' (Iredale et al., 1996)—are more likely to avail themselves of an Australian permanent residency, usually after a long application process. The way the process has been structured itself discriminates in favour of the people with higher 'human capital'.[3] The principle of greatest need is not entirely disregarded, of course, and the combination of the two principles—the greatest need and the resettlement potential—results in vastly heterogeneous refugee intakes. Therefore, alongside a minority of refugees who, due to harsh circumstances of their pre-Australian life, missed on formal schooling altogether and are virtually illiterate on arrival, all refugee communities (defined by the country of birth) comprise a considerable proportion of people with postschool qualifications.

During the 1990s and early 2000s, three regional groups, ex-Yugoslavs (mainly Bosnians), people from the Middle East (mainly Iraqis) and black Africans (from a number of countries) represented the overwhelming part of the Australian offshore humanitarian program (in some years as little as 6 per cent of the total annual immigration). The composition of our sample of respondents largely matches the composition of the recent Australian refugee intake. Since the late 1990s, black Africans have accounted for about 75 per cent of the annual refugee quota, with the largest number being from Sudan over the past seven years (Department of Immigration and Multicultural and Indigenous Affairs (DIMIA), 2005a). Due to their visibility and the 'novelty' value—until the recent refugee wave, the black African population in Australia was negligible—the Australian media regularly reports on their resettlement as well as associated 'integration problems' and the Australian public seems highly aware of their presence. A highly charged public debate on African refugees sprang up in 2005, and in October 2007 the immigration minister announced that, due to integration problems, the Sudanese refugee intake will be considerably reduced and will cease to comprise the majority of the humanitarian intake in incoming years (The Age, 2007). The vast cultural differences—perceived or real—and a perception that people who had been exposed to brutality and violence are themselves likely to be violent, created by media reports on African youth gangs, are mostly to blame for this policy shift.

Given the perception that refugees are, integration-wise, the most problematic immigration category and potentially a burden to the welfare system (Jupp, 2002), the Australian government provides a comprehensive settlement assistance to them through the Integrated Humanitarian Settlement Strategy (IHSS, see Department of Immigration and Citizenship (DIAC), 2007). The main aim is that it 'provides intensive settlement support to newly arrived humanitarian entrants' in order that they 'achieve self-sufficiency as soon as possible'. The latter primarily means achieving economic self-sufficiency, that is, employment. As stated on the government Web site, 'IHSS focuses on equipping entrants to gain access to mainstream services' (ibid.). Services within IHSS are offered to refugees for around six months, but 'may be extended for particularly vulnerable clients' (ibid.). Specifically, IHSS services comprise an initial needs assessment and referral to other service providers and mainstream agencies: on arrival accommodation for several months and follow-up assistance in finding a more permanent accommodation, usually in the private rental market but also in public housing; and short-term torture and trauma counselling services. Since the late 1990s, IHSS services have been contracted out to private agencies. Volunteer and community groups and charities have an important role within IHSS, primarily in assisting refugees in settling into the local community (DIAC, 2007).

As mentioned, a considerable proportion of the recent refugee intakes into Australia are people with skills and many have professional qualifications. For example, according to the 2001 census, all refugee communities, except Bosnians, had a higher proportion of people with higher qualifications than the Australia-born (the rate for the Australia-born was 18 per cent). At the same time they had much higher unemployment rate than the Australia-born (Table 5.1).

Previous research has shown that refugees resettled in advanced industrialised countries, Australia included, have much higher rates of unemployment than the resident population and that even when employed, they are likely to suffer forms of social exclusion such as underemployment, massive loss of their preflight occupational status, and isolation from the mainstream community, through working in immigrant employment niches.[4] This chapter focuses on the employment outcomes of recently arrived skilled refugees from three broadly conceived groups: ex-Yugoslavs (mainly Bosnians), black Africans (mainly Somalis and Ethiopians) and Middle Easterners (mainly Iraqis). These three refugee clusters make sense in terms of their differential visibility in the Australian context. In this chapter, visibility is understood

Table 5.1 Higher Qualifications and Unemployment for Selected Refugee Communities

Country of Origin	Higher Qualifications (%)	Unemployment Rate (%)
Bosnia	17.0	16.8
Sudan	26.2	27.6
Ethiopia	22.4	18.9
Eritrea	21.7	28.5
Somalia	13.6	46.8
Iraq	19.8	34.2

Source: 2001 Australian Census (DIMIA, 2005).

as a sum of those ethnic characteristics that make immigrants distinct in the Australian (Western, English-speaking) cultural context and among a predominantly white population. Visibility is a broader concept than 'race' or 'ethnicity', and at the same time less value-laden and burdened with various negative connotations. Visibility may comprise phenotypical features, dress and attire, accent, or any observable cultural difference. Visibility can be seen as the characteristic of a person that prompts the question 'Where do you come from?' In this way, the visibly different are being constructed as the 'Other', as outsiders who do not really belong. In contrast, members of the dominant group are construed as 'normal selves' and are therefore 'invisible', enjoying the privilege of anonymity and the absence of ethnic prejudice directed towards them.

The research project[5] this chapter reports on started with the assumption that visibility is especially problematic for recent immigrants, mainly to do with prejudice and discrimination in the labour market. In other words, in conjunction with refugee status, visibility tends to hamper employment outcomes, that is, a productive use of human capital. For a number of reasons, including visibility, the settlement of refugees is more difficult in virtually every aspect than the settlement of other two major immigration categories ('skilled' and 'family'), and the employment outcomes for refugees are considerably, and consistently, poorer than in the case of the other two categories (Richardson et al., 2004). In addition, the large Longitudinal Survey of Immigrants to Australia (LSIA) showed that a later cohort of refugees (Cohort 2: 1999–2000 arrivals) had worse labour market outcomes than Cohort 1 (1993–1995 arrivals) (Richardson et al., 2004: 1). As shown in this chapter, the reasons for generally poor refugee employment outcomes, and the specific cohort difference, may not be simply blamed on the lack of human capital—the usually quoted reason for poor labour market outcomes (Wooden, 1991)—but may have much to do with visibility and discrimination in the labour market.

Unlike in other areas of settlement, the Australian government does not provide special employment assistance to refugees, or to other non-English-speaking background immigrants, who are susceptible to labour market disadvantage (Commonwealth of Australia, 2007). In the late 1990s, the government-funded employment assistance was privatised and outsourced to agencies, collectively known as the 'Job Network' (JN). JN provides employment assistance to all Australian job seekers, including refugees (Australian Productivity Commission (APC), 2002). For the first six months, while they attend free English classes, refugees are exempt from seeking employment and accessing the employment services is optional. Afterwards, refugees have to start looking for jobs, just like any other Australian permanent resident or citizen, in order to continue receiving welfare payments. The current policy direction in government service provision, sometimes referred to as 'mainstreaming', as well as the privatisation of many government services, seems to be in conflict with the need for culturally specific services in a multicultural society. The Australian government's quest for financial efficiency, coupled with the significant retreat from the notion of multiculturalism, has signified an important shift away from what was previously known as 'multicultural services'.

The central body of data presented in this chapter was collected through a survey of 150 refugees permanently resettled in Perth, Western Australia. Questionnaire-based face-to-face interviews were conducted with fifty respondents from each of the three groups. Given the project's central emphasis on employment success, the interviewers (bilingual assistants) targeted skilled people of working age (18–65) with at least a working knowledge of English who were either employed or looking for work. Consequently, the sample was constructed (snowballed) to be purposive rather than representative of the refugee groups involved, and it consisted of respondents with the human capital (formal skills and English-language proficiency) considerably above average for the groups involved. Such a nonrandom sample means that the findings presented here are indicative for the (previously) middle-class people rather than easily generalised to the three refugee groups, which are, as mentioned, extremely heterogeneous in terms of socioeconomic background. The questionnaire consisted of three sections: demographic information, labour market experience and general satisfaction with settlement.

After the initial survey data analysis, nine in-depth follow-up interviews with key informants, mainly bilingual settlement workers of both genders, themselves refugees from the three groups, were conducted in order to further help interpret the survey findings, especially with regard to cross-cultural issues involved. This chapter also briefly reports on the qualitative data set gleaned from the interviews with forty Australian employers, which focused on their experience of employing the 'visibly different', and specifically people from the three refugee groups. The sample of employers comprised a broad cross section of industries, organisations (eight small, eight medium and twenty-four large companies), and managerial profiles. Many of the smaller organisations we approached declined to participate, citing reasons such as lack of time and lack of experience with migrant employees. People who represented employers in interviews were twenty-five women and fifteen men, with an average age of forty-one. Of these, twenty-two were human-resource managers who facilitated the recruitment and hiring process, and the rest were people on various managerial positions. Only three interviewees were from non-English-speaking backgrounds, though more were migrants (mainly from the UK). Twenty out of forty interviewees had tertiary qualifications.

In terms of human capital, our refugee sample represented the upper end of the refugee communities surveyed (Table 5.2). Table 5.2 shows that all respondents in our sample reported having at least twelve years of schooling (which in ex-Yugoslavia often means having a vocational qualification), and a large proportion of each subsample had a university or even a postgraduate degree. These highly skilled refugees could therefore be expected to have

Table 5.2 Highest Education Level (%; N = 150)

	Ex-Yugoslav	African	Middle Eastern	Total
High School 12 years	28.0	10.0	2.0	13.3
TAFE* diploma	38.0	22.0	32.0	30.7
University degree	26.0	42.0	46.0	38.0
Postgrad qualification	6.0	26.0	20.0	17.3
No response	2.0	0.0	0.0	0.7

*'Technical and further education', i.e. the nonuniversity tertiary education sector.

patterns of employment that closely resemble immigrants arriving through the skilled stream. In addition, a large majority of our respondents had lived in Australia for five years or more: 96 per cent of ex-Yugoslavs, 80 per cent of Africans and 58 per cent of respondents from the Middle East. The African and Middle Eastern participants were younger and predominantly male, and the ex-Yugoslav subsample was predominantly female and somewhat older (Table 5.3). In addition, the African and Middle Eastern subsample had higher English proficiency (Table 5.4). These obvious labour market advantages did not, however, translate to either better employment status (see Table 5.5) or a higher average income (Table 5.6).

The ex-Yugoslav subsample was older (on average ≈ 44 years) than African and Middle Eastern subsamples (≈ 37 and ≈ 38 years, respectively) and

Table 5.3 Sample Characteristics (N = 150)

	Range	Ex-Yugoslavs Mean (SD)	Africans Mean (SD)	Middle Eastern Mean (SD)	Total Mean (SD)
Age[A]	1–5	3.10 (.64)	2.64 (.76)	2.70 (.58)	2.99 (.77)
Gender		F = 58%	F = 28%	F = 28%	F = 38%
Education[B]	1–4	2.80 (.57)	3.16 (.58)	3.18 (.44)	3.05 (.56)
Length of residence[C]	years	7.78 (2.10)	7.24 (3.04)	6.54 (4.30)	7.19 (3.15)

Notes: [A]The age scale: 1 = under 20; 2 = 21–35; 3 = 36–50; 4 = 51–65; 5 = over 66.

[B]The education level scale: 1 ≤ 10 years; 2 = trade or 12 years; 3 = diploma or degree; 4 = postgraduate qualification.

[C]Medians (years): ex-Yugoslavs *Mdn* = 8.00; Africans *Mdn* = 7.00; Middle Easterners *Mdn* = 5.00.

predominately female. The African and Middle Eastern subsamples were almost identical in terms of age and gender composition and very similar in terms of education and length of residence. A large majority of Africans and Middle Easterners had university or even postgraduate degrees (68 and 66 per cent, respectively), while ex-Yugoslavs were somewhat less highly educated (32 per cent with university degrees) and with lower self-reported English proficiency. More details on English proficiency are presented in Table 5.4.

Table 5.5 shows current employment status of our participants. The Middle Eastern respondents have the worst employment status: the highest unemployment rate, the lowest proportion of people in full-time jobs and the lowest proportion of people working part time, more than eleven hours a week. This is in spite of their highest education score among the three groups (Table 5.3) and a high self-reported English proficiency (Table 5.4). Ex-Yugoslavs have convincingly better employment status than the other two groups.

Table 5.6 shows that ex-Yugoslavs have convincingly higher average fortnightly income than the other two groups. The difference is even larger

Table 5.4 Current English Proficiency (Self-Assessed) (%; N = 150)

	Ex-Yugoslav %	African %	Middle Eastern %	Total %
Spoken:				
No English	0.0	0.0	0.0	0.0
Basic	0.0	2.0	0.0	0.7
Good	32.0	30.0	12.0	24.7
Very good	38.0	38.0	66.0	47.3
Fluent	28.0	30.0	22.0	26.7
No response	2.0	0.0	0.0	0.7
Written:				
No English	0.0	0.0	0.0	0.0
Basic	6.0	2.0	4.0	4.0
Good	40.0	30.0	24.0	31.3
Very good	32.0	40.0	52.0	41.3
Fluent	20.0	28.0	20.0	22.7
No response	2.0	0.0	0.0	0.7

Table 5.5 Current Employment Status (%; N = 150)

	Ex-Yugoslav %	African %	Middle Eastern %	Total %
Unemployed	14.0	32.0	38.0	28.0
Less that 10 hrs p/w	6.0	2.0	4.0	4.0
11–20 hrs p/w	18.0	10.0	10.0	12.7
21–30 hours p/w	8.0	18.0	6.0	10.7
Employed full time	52.0	36.0	40.0	42.7
No response	2.0	2.0	2.0	2.0

when household income is taken into account, probably reflecting the cultural standard of both spouses in the household working outside the home, whereas many African and Middle Eastern women stayed at home caring for their families (due to a cultural standard of the male breadwinner, but also to the fact that these were significantly larger and younger families than the ex-Yugoslav ones). The higher individual income of ex-Yugoslavs may be partly due to their somewhat longer residence in Australia, but on the other hand it contradicts the apparent labour market disadvantage of being female and older (in comparison to younger and predominantly male African and Middle Eastern participants).

Over the past fifteen years, Australia has experienced steady economic growth and prosperity, with unemployment levels reported to be at their lowest in thirty years—currently under 5 per cent across the nation. Despite this, a majority of our respondents, and especially those from the Middle East, reported having difficulties finding work (Table 5.7).

Respondents were also asked about discrimination in the job market (Table 5.8).

The degree of reported discrimination—when different grounds of discrimination were included—was the highest for the Middle Eastern group, which is consistent with their lowest reported income and the worst employment status among the three groups. While the Africans, perhaps surprisingly, reported the lowest level of difficulty finding work (but still no less than 50 per cent), they provided many written comments expressing their

Table 5.6 Average Net Fortnightly Income (A\$; N = 150)

	Ex-Yugoslav	African	Middle Eastern	Total
Household	2257.6	1521.2	1238.5	1650.9
Individual	913.2	799.2	704.7	801.2

Table 5.7 Experienced Difficulties Finding Work (%; N = 150)

	Ex-Yugoslav	African	Middle Eastern	Total
No	42.0	48.0	20.0	36.7
Yes	58.0	50.0	78.0	62.0
No response		2.0	2.0	1.3

frustration at their lack of success in the job market and having to take jobs below their formal qualifications. Many Middle Eastern and African Muslim respondents commented on the impact of the 9/11 terrorist attack on their lives and their experience of discrimination because of their appearance, accent and Muslim name. Five respondents volunteered stories about being disadvantaged by their 'unpronounceable' and/or Muslim names, and stated that when they either changed their name or used only an initial on their resume, they received more positive responses from employers (see also Human Rights and Equal Opportunity Commission (HREOC), 2004).

The attribution of different types of discrimination varied significantly across groups. Language ability was much less of an issue for Africans and Middle Easterners, which is not surprising given their reports on English proficiency. Accent was seen as more of an issue for ex-Yugoslavs, although for Africans, accent (sometimes referred to as 'African English'), together with appearance, was a likely cause of discrimination. For Middle Easterners, accent, name and appearance fared equally. The 22 per cent of Africans who felt they had been discriminated against in the job market due to

Table 5.8 Experience of Discrimination in the Labour Market (%; N = 150)

	Ex-Yugoslav %	African %	Middle Eastern %	Total %
No	50.0	52.0	42.0	48.0
Yes	48.0	40.0	52.0	46.7
Basis of discrimination*				
Language ability	42.0	8.0	16.0	22.0
Accent	40.0	28.0	28.0	32.0
Name	18.0	24.0	28.0	23.3
Appearance	8.0	28.0	28.0	21.3
Religious customs	0.0	22.0	12.0	11.3

*Participants could select more than one category.

religious customs reflected the large proportion of Somali Muslims among our participants. The language ability, which is the only basis of job-market disadvantage pertaining to a job-relevant characteristic (together with foreign accent, in case it is as strong as to hamper communication), was quoted as a basis for discrimination by a large proportion of ex-Yugoslavs (42 per cent, as compared to 8 and 16 per cent, respectively, for the other two groups). This is also the main form of visible (or, rather, 'audible') difference for ex-Yugoslavs in the Australian context, as opposed to phenotypical differences. This is consistent with their lower self-reported English proficiency, but reporting so much discrimination on the basis of language ability and accent (82 per cent when combined) may also reflect a relative absence of other perceived causes of discrimination such as appearance and religion, which are often quoted by Africans and Middle Easterners. The fact that 48 per cent of ex-Yugoslavs claimed that they had been discriminated against in the labour market may not seem consistent with their relatively good employment outcomes and income (as compared with other two groups), but it is consistent with the fact that no less than 80 per cent of ex-Yugoslavs had jobs below their qualifications, compared with 44 per cent of Africans and 24 per cent of Middle Easterners. Given the fact that women formed a larger proportion of the ex-Yugoslavian sample, the latter difference may reflect a higher willingness of women to accept a job below their qualifications in order to help out their family finances, rather than remaining unemployed, or updating their professional qualifications in Australia, in the hope of regaining the previous occupational status, which may have been the case with a number of the highly educated and younger African and Middle Eastern men in our samples.

Participants were also asked to identify a number of structural barriers to employment, as shown in Table 5.9.

A significant proportion of participants from all three groups quoted a lack of Australian work experience and the related inability to provide Australian references as barriers to securing adequate employment. Even when formal qualifications were recognised, this did not seem to be of much worth without local experience. Almost all respondents who provided written comments mentioned the issue of local experience. A minority saw this as understandable and justified, while a significant majority described this as a 'Catch 22' situation and therefore unfair. Many respondents saw the lack of recognition of overseas experience as a case of structural discrimination, an 'excuse' to deny them adequate employment, while perceiving the real reasons as lying in non-work-essential characteristics such as name, appearance, and the religious, cultural and ethnic differences they imply. Table 5.10 reports on the job satisfaction of our respondents.

The relatively high job satisfaction among ex-Yugoslavs, in spite of the lowering of the occupational status for 80 per cent of respondents, may be attributed to the fact that they earned a satisfactory income from the job (see Table 5.6). As our previous research among Bosnian refugees shows,

Table 5.9 Barriers in Securing Adequate Employment* (%; N = 150)

	Ex-Yugoslav %	African %	Middle Eastern %	Total %
Problems getting qualifications recognised	32.0	42.0	20.0	31.3
Lack of Australian work experience	68.0	58.0	60.0	62.0
Lack of Australian referees	54.0	48.0	20.0	40.7
Lack of work experience due to life in refugee camps	18.0	4.0	8.0	10.0
Breaks in working life	18.0	16.0	8.0	14.0
Difficulties getting promoted	30.0	2.0	8.0	13.3
Necessity of having a car	34.0	26.0	8.0	22.7
Other	10.0	2.0	10.0	7.3

*Respondents could select more than one category.

material prosperity with its culturally appropriate symbols (e.g. a large, well-furnished house) was a primary goal for many people who had lost everything when they fled their homeland. Therefore, a reasonable wage, in a situation where family earnings were pooled in order to arrive at culturally valued material goals—and the status they bring within the community—may have been more highly valued than the job status itself (Colic-Peisker and Tilbury, 2003).

In the Perth metropolitan area, where our research was conducted, refugees were concentrated in low-skill service 'niches' such as cleaning, aged-care, transport (especially taxi driving), food-processing and security and building industries. The fact that highly skilled non-English-speaking-background immigrants drive taxis in large Australian cities is these days

Table 5.10 Satisfaction with Current Job (%; N = 150)

	Ex-Yugoslav %	African %	Middle Eastern %	Total %
Entirely	30.0	10.0	18.0	19.3
Mostly	32.0	26.0	26.0	28.0
Somewhat	18.0	20.0	12.0	16.7
Not at all	16.0	10.0	10.0	12.0

a common knowledge among cab users (Constable et al., 2004). Many low-skill niche jobs take place in the context of the informal economy. Needless to say, these are poorly paid and insecure jobs, although some ethnic entrepreneurs make envious profits by employing—and exploiting— their recently arrived compatriots. We found Africans concentrated in food processing, the security industry and aged-care jobs. Aged care is a booming industry, but not one that offers attractive jobs and therefore left to certain categories of immigrants. Our respondents described the security industry, where a number of Africans had also found jobs, as 'dangerous'.

The concept of ethnic or immigrant niche has been used in literature to describe the spatial concentrations of immigrant-owned businesses, predominately employing their coethnics and usually also serving 'ethnic' clientele (Barth, 1994). 'Ethnic niche' can also denote an intense presence of particular ethnic groups in certain industries due to their particular culturally determined skills or practices (Bonacich, 1979). This chapter reports on a different type of employment niche, signifying the unintended concentration of refugees, in the role of employees, in certain unattractive industries as a consequence of labour market constraints placed on them, rather than their choice or an entrepreneurial spirit.

Relying on ethnic community networks in securing a job reinforces directing people into certain employment niches. Through sharing information on job vacancies, but also through the community perceptions of what can be expected and achieved in the labour market, job seekers from certain communities end up concentrated in certain industries. Fifty-nine per cent of our respondents reported using ethnic community networks in looking for a job, while 47 per cent of people actually found jobs through these networks. The reliance on ethnic networks and finding jobs through them was the highest among ex-Yugoslavs, 72 and 66 per cent, respectively. 'Other networks' (in the wider community) were only used by 18 per cent of respondents across the three groups, and 11 per cent secured a job through them. Ethnic-path integration created a high concentration of ex-Yugoslav refugees in the cleaning industry: most of the 88 per cent of our sample who entered the Australian job market as unskilled workers first worked as cleaners.

A human-resource manager of a large cleaning company described the process that reinforces the existence of ethnic employment niches:

> I've found through experience [that] a lot of it is a word of mouth. When you've got a thousand staff, they become your advertising network themselves. We fill a lot of our positions through people who know people. . . . [If] they come at the door looking for work, they're generally associated with somebody who's already working for us.

Ex-Yugoslavs arrived in large numbers during a relatively short period in the mid-1990s and formed tight-knit communities. They also found large communities of longer term migrants of their ethnic backgrounds in Australian cities, as well as 'ethnic businesses' that could provide jobs. This was not the case with the other two groups that largely fall into the category of 'new and emerging'[6] communities.

Ethnic firms tend to be small businesses in retail and construction, which as a rule employ labourers, tradespeople, and only occasionally office workers and professionals. Therefore, they present little employment opportunity for the highly skilled (see also Lamba, 2003:48). Ethnic entrepreneurs often consciously employ newly arrived co-ethnics expecting them to be cheap and pliable labour. A manager of a Perth 'ethnic' construction firm told us:

> Immigrants of Croatian and Bosnian background often look for jobs [here]. They hear about the firm from others in the community, then ring us, and we ask them to fill in the recruitment form . . . The owner considers the same-background immigrants to be hard workers who rarely complain about working conditions as long as they feel they can earn enough. The conditions are harsh [in the North of Western Australia] . . . outdoors in the dust, in 50 degrees temperature . . .

Therefore, in the context of job search, ethnic community networks are enabling as well as constraining. Community cohesion and conformism may create a pressure towards 'downwards adjustment' for better-educated people (Bloch, 2002; Colic-Peisker and Walker, 2003). Another, subtler mechanism supporting downwards occupational mobility is a community perception that, due to mainstream prejudice and discrimination, looking for a good job is an uphill battle and a highly insecure investment of time and effort (cf. Fugazza, 2003). Consequently, accepting a low-skill job, at least as a start, is considered a more rational course of action. A migrant employment officer described the situation among recently arrived Africans:

> There's quite a widespread perception among Africans that discrimination is widespread . . . 100 per cent of people we dealt with have at some stage identified discrimination, they all experienced discrimination one way or another . . . specifically they would say 'it's all about the colour of my skin' [or] 'the way I speak' . . . These stories go around and are taken on board . . . People I worked with were entrenched into the negative [but] eventually when we moved on and got them a work experience they changed a bit . . .

Employment in the niches of the secondary labour market, often meant as a temporary solution, easily becomes a trap for people with professional skills: after a certain amount of time elapses, skills may become dated and the gap in the professional CV may represent another obstacle in finding an appropriate job. In addition, the low labour market location prevents the creation of 'weak ties'—the networks in the wider community, outside the 'strong ties' of extended family and ethnic community—that are highly instrumental in securing a professional job (Granovetter, 1973).

International research shows that 'informal' racism expressed through the prejudicial behaviour of employers represents another mechanism that reinforces channelling refugees into low-skilled labour market niches. Shih (2002) showed that hiring decisions of American employers are influenced by racial stereotyping and not made on purely economic grounds. In our sample of Australian employers, those outside the identified migrant employment 'niches' had little or no experience with employing the 'visibly different' from the three target groups. Almost everyone emphasised the job seekers' skills as the element that determined the hiring decision, but when exposed to more intricate questions and examples, stereotyping of ethnic groups did occur, and in several cases employers in the private sector (mainly older male interviewees who held power in the firm) openly expressed prejudices. Nonessential characteristics of an employee, sometimes called 'soft skills', were alluded to in the often-emphasised idea of 'organisational fit' and 'personality match' (see Tilbury and Colic-Peisker, 2006, for more details). Employers also tended to rely on the reputation of ethnic communities, and if a group was considered 'hardworking'—e.g. this is a reputation that ex-Yugoslavs have had in Australia for decades—community clustering at work was encouraged. Stereotyping can work against certain groups, and indeed it does in the case of Africans and Middle Easterners, although for different reasons. In addition, refugees are perceived as the lowest class of immigrants, people coming from global trouble spots and therefore unfavourably stereotyped in general. Given the fact—in the eyes of many—that they were 'saved' by being accepted in Australia, they are 'naturally' expected to accept jobs that locals do not want. We learned that the Department of Immigration received calls from desperate employers looking for 'refugees' for the hard-to-fill job vacancies (in meat works and aged care, for example). Most employers were careful not to express any prejudice against the visibly different, and the poor employment outcomes of highly skilled refugees were blamed on various objective market conditions rather than on everyday racism and discrimination.

CONCLUSION

The education and middle-class background of many recent refugees may be perceived by Australian authorities, concerned about 'social cohesion', as a guarantee that these people are, to a degree at least, 'people like us', socialised into urban Western cultural practices and therefore able to successfully integrate in the Australian social context. However, their socioeconomic endowments, and especially professional skills, are often wasted in Australia through high levels of unemployment and, for those employed, a massive loss of occupational status. If the findings of this research project had to be expressed in one sentence, this would be: labour market outcomes of refugees equal human capital minus visibility. Consequently, ex-Yugoslavs scored better than the other two groups—black Africans and Middle Easterners—on most employment indicators, in spite of having a lower education profile and poorer English on average.

Our findings contradict the view that there are 'no separate labour market segments for Non-English Speaking Background (NESB) immigrants' (Adhikari, 1999:203) and that 'labour market is blind to ethnicity' (Evans and Kelly, 1991). In turn, they confirm the view that there are 'high and sustained levels of occupational segregation, particularly amongst migrants from NES countries' (Wood, 1990: 2; see also Ho and Alcorso, 2004; Collins, 1991). It seems that a segmented labour market, where racially and culturally visible migrants, and refugees in particular, despite their skills levels, are allocated unattractive jobs in the secondary labour market, has persisted into the twenty-first century (more details in Colic-Peisker and Tilbury, 2006). The poor labour market outcomes for skilled refugees persist at the time of the thirty-year low unemployment in Australia and amidst serious skill shortages in many industries.

Prejudice against people associated with refugee-generating countries (even if they did not arrive on refugee visas) and a discrimination on the basis of race, religion and ethnic origin plays a role in creating unsatisfactory employment outcomes. It is notoriously hard to establish, however, how big a role prejudice and discrimination play in hampering labour market success. From our data it is obvious that, for example, employers tend to pay much attention to so-called 'soft skills' such as Australian cultural knowledge and an appropriate 'tearoom mentality' (e.g. certain sense of humour and general 'attitude') when hiring their employees. In consequence, this criterion leads to discrimination against the 'culturally different', regardless of their skills and ability to do the job well. The cultural difference—which translates as a cultural inadequacy of the non-English-background minorities—is often simply extrapolated from the visible difference (physical features or dress), which indeed is the very essence of racism, or from the negative mainstream view of the country and culture refugees come from. Among employers, structural discrimination on the basis of cultural difference is perceived and interpreted as a 'just' market mechanism that allocates those with individual limitations (e.g. foreign accent or religious observance) where they rightfully belong—to the lower rungs of the labour market.

If the problem of disadvantage and discrimination against the visibly different refugees in the Australian labour market is not addressed by policymakers, there is a possibility that, as a result, the 'new and emerging' refugees communities—and specifically African and Middle Eastern—may develop into marginalised minorities riddled by social problems. A similar process has affected Vietnamese and Lebanese refugee communities that arrived in Australia in the 1970s–1980s.

What can the policymakers do? Apparently, a national system of qualifications recognition needs to be more sensitive and flexible, and needs to include greater regulation of professional organisations and registration boards, as nonrecognition or partial recognition of qualification is a real as well as symbolic obstacle to labour market success. That is, while formal qualification recognition does not in any way guarantee a good job—this is entirely left to labour market forces—the nonrecognition may reinforce hopelessness in recent arrival and prejudices of employers against certain national groups. We may ask, given that some professional organisations (e.g. Australian Medical Association) make their elaborate exams necessary for practicing the profession in Australia, and given that the central government's recognition of qualifications does not in any way guarantee employment, what indeed is the point of qualification (non)recognition? Could not then market forces alone make an immigrant go back to school in order to update his/her qualifications, rather than being humiliated through an official government declaration that one's diploma or degree is considered worthless in Australia?

On the other hand, employment agencies could do more to provide targeted professional training through work-experience placements. 'Work placements' refer to voluntary work in skilled and professional occupations that may, and often does, lead to paid employment, especially in the areas of skill shortages. Our participants (a focus group of employment workers in migrant resource centres) reported that the work placements are difficult to organise, however, because many employers view that as liability rather than an asset. In cases when refugees do get a work placement, in a great majority of cases this leads to paid-employment opportunity. It not only provides a much-needed introduction into the Australian work culture and style in a given profession but also secures a necessary Australian work experience and Australian referees.

On a broader plane, the Australian government could conduct public-awareness campaigns to make employers aware of the benefits of workplace diversity for their work environments and productivity, so 'market imperatives' are no longer used as excuses for the labour market exclusion of the visibly different but rather work in favour of the diverse workforce. Employers also need to be educated about what constitutes discrimination because sometimes they are genuinely unaware they engage in discriminatory practices. In this respect, political leadership is also necessary to turn around the climate of hostility towards those who are specific visible minorities—Muslims in particular, but also black Africans. The government should also ensure that antidiscrimination agencies increase their public profile and are more proactive in encouraging people to report their grievances. Links between those providing settlement services to the visibly different and HREOC (Human Rights and Equal Opportunity Commission) may result in more cases coming to light and have a preventative effect. Our research shows that many people, although they feel they have been discriminated against, see no point in reporting their grievances.

It should be kept in mind that this chapter reports on the relatively early experiences of the recently arrived refugees and that two-thirds of our respondents belong to the 'new and emerging' communities (Iraqis and black Africans). Once these communities are more established and some of their members, in spite of considerable obstacles, secure professional jobs, the mainstream perception of these ethnic groups will inevitably improve, as it happened in the past to many other groups, Italians and Greeks being the most notable examples. However, these processes cannot be simply left to time and 'market forces' and government initiatives are necessary. The image of a refugee in the mainstream community needs to be improved: a general medicalisation of the refugee experience places an excessive emphasis on refugee trauma, illness and mental-health intervention (often culturally insensitive and inappropriate) and creates an image of a refugee as a needy, disabled and helpless individual. The medical emphasis diminishes a focus on social integration and within it to the crucial issue of employment as a most significant basis for social inclusion. We therefore advocate a stronger emphasis on social inclusion, especially finding employment and establishing wider social networks, as a prevention of social exclusion and marginalisation. The latter, as is known from the mainstream social psychological research, can in itself cause health and mental-health problems.

NOTES

1. 'Offshore' refers to the planned humanitarian intake of people who arrive in Australia on permanent residency visas while 'onshore' refers to asylum seekers who enter the country 'unauthorised' (usually by boat via Indonesia) in order to seek residency on arrival (see http://www.immi.gov.au/refugee/migrating/index.htm). The latter category comprises a very small proportion of the total humanitarian immigration stream (up to 1/10) and is strongly discouraged through various legal regulations and policy measures (e.g. mandatory detention of asylum seekers; excision of northern Australian regions and islands for immigration purposes; only granting a three-year temporary protection if asylum seekers are found to be 'genuine refugees'. This chapter only reports on the 'offshore', permanently protected refugees.
2. The offshore humanitarian program comprises several categories of entrants, but all of them are, for brevity, referred to as 'refugees' in this chapter.
3. In this chapter the concept of 'human capital' is used as a shorthand for all those skills and competencies that enable a person to secure a good job, that is, consistent with the economic notion of 'capital', to reap higher returns in the labour market. Apart from human capital, there are indications that being a Christian (most of the recent African refugee intake) or a white European (the large Bosnian intake in the 1990s) also improves one's chances to be selected for the Australian 'offshore resettlement'.
4. Valtonen (1999, 2004) analyzed this situation in Finland and the Netherlands; Lamba (2003) in Canada; Rydgren (2004) in Sweden; and Colic-Peisker and Tilbury (2006) in Australia.
5. The project 'Refugees and employment: the effect of visible difference on discrimination' was funded for three years (2004–06) by the Australian Research Council Discovery Project Grant DP0450306. Chief investigators were Val Colic-Peisker, Farida Tilbury and Nonja Peters.
6. This is an official policy term.

6   Migrants' Paths in the Italian Labour Market and in the Migrant Regulatory Frameworks
Precariousness as a Constant Factor

Giovanni Mottura and Matteo Rinaldini

Even though every development of Italian migration policy has included aspects of discontinuity, often presented as remedies for the failure of previous laws, it is possible to trace a strong line of continuity, both in terms of its policy elaboration and of its effectiveness. Recent studies on the origins of Italian migration policies have pointed to a steady presence of a demand for 'pragmatic functionalism' from employers and the persistence of a powerful alignment between this and attitudes of solidarity, represented by the advocacy coalition (Zincone, 2006). Other studies have highlighted the presence of a high level of discretionary power that has distinguished all Italian migration policy (Triandafyllidou, 2003; Dell'Olio, 2004). This chapter will try to identify some common themes across different Italian migration policies, to relate them to the way in which migrants have been inserted into the labour market. The aim is not to demonstrate that migration policies are the only factor determining the migrant labour market experience. Nevertheless, it seems legitimate to argue that, on the one hand migrant policies have nourished chronic problems within the Italian labour market (such as the widespread and structural persistency of irregular employment) and on the other hand have played a role in encouraging transformations in labour conditions (as evidenced by the increasing demand for flexible work conditions) by ignoring the specific situation of migrant workers. We then move our focus on two recurrent factors in migration policies: the extended use of mass legalizations as the principal device for regularizing migrants' residency; and the continuing development, within the legal dimension, of a hierarchical integration model for migrants. In a second section of the chapter, based on some results of a survey, conducted in 2003, on 1,654 migrant workers living in a prosperous region in the North East, we highlight the specifics of migrants' working conditions. Finally, in the last section we explore the relationship between the labour market and migration, using the Marxian theory of *a reserve army of labour*, which we view, in its most authentic interpretation, as a core or functional element (rather than a marginal or dysfunctional one) in the processes of the reproduction of capital and at the same time as a component of the supply of labour, not solely reducible to the categories of unemployment, secondary labour market or marginal mass, as it has often been characterised.

THE REGULATORY FRAMEWORK IN ITALY

From the beginning of migration to Italy, mass legalization has been the main device for regularizing migrants' residency rights. The first mass legalization dates back to the time when Italy did not have any legislation governing migration: in 1982, 15,000 residency permits were issued, almost all of them for domestic work. Four years later, in 1986, the first legislative measure was adopted to regulate the entry of migrants into Italy (Foschi Law 943/1986), and at the same time a second mass legalization was initiated. The regulatory model for entry, based on the notion of the nonavailability of local labour, was a failure, but mass legalization had regularized 118,349 migrants. A few years later, in response to increasing migratory pressures, the Martelli Law (39/1990) introduced a regulatory model based on the establishment of annual quotas for entry. Notwithstanding their planning, the annual quota system was not achieved. The mass legalization in 1990, launched by the Martelli Law itself, regularised 234,841 migrants, and as a consequence it induced the government to postpone the introduction of further legislation. In fact, the principles adopted for the 1990 mass legalization were relatively straightforward. Employed and self-employed migrants could obtain residency permits, as could the unemployed (although, in the case of the later, they had to register with the employment office). This type of permissive mass legalization is unlikely to reoccur. Contemporaneously, Italy agreed to join the Shenghen group, an event revolutionizing the movements of people into Italy, by reinforcing external borders and by imposing undocumented status on all who enter the national territory without permission. In this way the conditions had been created which forced the state to turn periodically to mass regularisations, so as to regularize the position of those migrants already in Italy. The inability of the state to set quotas, together with the strengthening of external borders, determined that in the following years there was a limit on legal entry and an inevitable increase in irregular migration. In 1995, another mass legalization took place. Its characteristics were different from the previous ones: residency could be regularized only for those in regular employment. This principle would be maintained in subsequent mass regularisations.

Notwithstanding, the number of migrants regularised in 1995 was 258,501. In 1998 a new migration law was approved: the Turco-Napolitano 40/1998 Law, subsequently amended by the Testo Unico sull'Immigrazione. This was followed, in 1998, by another mass regularisation of 250,747 migrants. Nevertheless the new law created a new regulatory model of entry for regular migration, and for the first time (with the exception of two small quotas planned in 1996 and in 1997) quota planning began to work. The regulatory model provided for the annual planning of quotas, which would be the main entry route for work. The setting of these quotas was to have been based on the periodical survey of labour market statistics and on bilateral agreements with migrants' countries of origin. What is interesting is the relatively limited use recourse to the quota numbers from 1998 onwards. In 1998 and 1999 the quotas, still in an experimental phase, were 58,000 in each year but only half of these were taken up. In 2000, 83,000 work permits were issued. However, 20,000 of them were short-term, nonrenewable seasonal work permits. In 2001 the number of work permits increased to 89,400, 39,400 of which were for seasonal workers. In 2002 the number of work permits decreased to 79,400; however, seasonal permits increased to 50,000. In 2003 79,400 work permits were issued, but seasonal work permits increased to the unprecedented level of 68,000. In other words, the quota for nonseasonal work permits (which allow migrants access to more stable and longer-term employment) fell continuously. Paradoxically, by 2003 there were fewer such permits than in 1996 (10,000 compared to 23,000). These characteristics of planning of migrant flows, hand in hand with the further strengthening of borders under the 286/98 Law, help explain why, in 2002, the largest mass regularisation (of 646,000 migrants—the Bossi-Fini Law) ever seen in Europe occurred. Thus, prior to 2003, the main model for Italian migration policy was based on the posthumous regularization of migrants' presence. These regularizations constituted the bulk of the work permits issued. Mass regularisations represented a 'necessary defection' that enabled migrants already in the country to emerge from irregularity (Einaudi, 2007). Taking into consideration that these happened every three to four years, it is possible to conclude that hundreds of thousands of migrants had been exposed to economical and social marginalization dynamics, until they obtained the necessary legal status to allow them to begin their integration into the labour market and into the Italian society (Mottura, 2003).

From 2004, with enlargement bringing about the EU25, there was a deep change in the setting of work quotas. Italy (as was the case for almost every old member country of the EU) put into force restrictions on the free movement of new member-state citizens, but this was done in a moderate manner. In 2004 a quota of 79,500 work permits for non-EU migrants was set (25,000 of which were for seasonal jobs) with an additional quota of 36,000 work permits for new member-state citizens. Similarly, in 2005 a quota of 99,500 work permits for non-EU migrants was set (of which 45,000 were for seasonal jobs) with an additional quota of 79,500 work permits for new member-state citizens. In 2006 a quota of 170,000 work permits was set (50,000 of which were seasonal jobs) for non-EU migrants with an additional quota of 170,000 work permits for new member-state citizens. In total, 340,000 work permits were issued during 2006. The reversal of the previous trend, however, resulted in a meeting of new dynamics and old procedures. First, some work permits were issued for specific work sectors, and the request could be in relation to a specific job (for instance, in 2006 within the quota for non-EU migrants, 45,000 work permits were issued only for domestic work). Second, the reduction of work permits for seasonal jobs within the whole non-EU quota was based on a forecast that seasonal manpower would have been found from among new member-state citizens. Indeed, subsequently it was ascertained that the actual number of new member-state entries was lower than the fixed quota, while work permits for non-EU seasonal workers started to increase again. Third, it was now clear that it was undocumented and overstayer migrants who were taking advantage of the quotas and not those currently living abroad. Actually the quota system, far from being an alternative model to that of mass regularisation, reproduced the same social dynamics.

At the beginning of 2000, just as the quota system was being consolidated (with all of its ambiguities as mentioned earlier), another characteristic of Italian immigration policy was established: the development of a legal hierarchical integration model. As already mentioned, Italy's joining the Shenghen group countries in 1990 had led to the extension of undocumented status to all migrants from non-Shenghen areas entering Italy without a permit. The distinction between regularity and irregularity of entry (determining whether legalisation might occur) became stronger. However, what had really happened was that the legal processes for integration had been substantially transformed. In 1992 a legislative measure had been adopted restricting the possibility of obtaining citizenship, even for those migrants born in Italy, unless they had been legally resident, without a break, from birth to the age of 18. The previous legislation required simply that the individual was aged 21 or over and was resident at the date of application. Moreover, for those not born in Italy, the new law changed the number of years of residence necessary to obtain citizenship. It penalized non-EU citizens who had to have at least ten years of residence to apply for citizenship, which is not granted automatically. There is a high level of state discretion, to the point of even taking the individual's personal income into account. Therefore, from the early 90s there have been two opposite positions in relation to integration: a lower pole, of irregular status, resulting in a complete exclusion from citizenship-based rights; and a higher pole, conferring citizenship status and offering equal legal rights to migrants. In the years following 1992, the 40/98 and 189/2002 laws were aimed at regulating the legal position of those occupying the space between these poles. In 1998, the 40/98 Turco-Napolitano Law established a real integration model on a legal basis. First, in line with previous legislation, it provided that different kinds of residency permits had to correspond to the reasons for entry. These different permits in turn provided differential and unequal rights. Moreover, in relation to entry for work reasons, there was a further differentiation: the type of work permit depended indeed on the nature of the job (duration, work contract, etc.), and only some work permits gave a right of renewal (or of conversion to another kind of residency permit).

Second, the law provided that the process of permit renewal was not automatic but had to be linked to the possession of specific requirements (regular residence, a regular work contract, sufficient income, no criminal record). Third, the legislation to ratify an EU directive, provided for a *Carta di Soggiorno* (long-term residence permit), a permission to remain in Italy without limit that could be obtained by migrants after five years of regular stay in the country. However, the issue and maintenance of the *Carta di Soggiorno* was subject to specific economic and penal requirements that had to be met by migrants. Contemporaneous to the creation of an integration model for regular migrants, the devices to tackle illegal and irregular migration were strengthened (for instance, by building structures known as C.P.T, 'Temporary Permanence Centres' for the identification and holding of migrants without documents). In other words, the Turco-Napolitano law created an integration model characterized by a strong tendency to *civic stratification* (Morris, 2001), based on different statuses (degrees of citizenship), never definitively acquired and in which regular employment constituted the main *rule of transition* between the statuses (Baubock, 1991; Morris, 2003). Some critical voices have pointed to the risks of that integration model. In particular they underlined the fact that, in a context characterized by an increasingly flexible labour market and by the development of new typologies of work contracts (nonstandard), the employment contract as the main determinant of status transition exposed migrants to the risk of loss of any previously acquired status (Mezzadra, 2001). It is also true that in the period following the 40/98 Law, given its mode of application, the resulting risks turned out to be relatively low (Pugliese, 2002). However, the changes brought by the Bossi-Fini Law (189/2002), four years later, demonstrated the ambiguous nature of the integration model it had created. The Law of 189/2002 neither modified nor distorted the previous legislation in a consistent manner. However it had some novel features, changing the integration models, and the result was that those elements that were most indistinct were accentuated. First, the concept of a *contract to stay* was introduced, that is, an agreement between the migrant worker and the employer necessary to obtain or to renew the residence permit. The maximum duration of the contract to stay was two years in the case of an open-ended work contract, one year in the case of a temporary contract and nine months in the case of a seasonal job. Since the residence permit was dependent on the contract to stay, its expiry dates were automatically linked. Consequently, the length of the residence permit was always dependent on the nature of the work contract, in contrast to the previous legislation, whereby a renewal of the work contract meant that the residence permit would automatically be renewed. Second, the new law lowered to six months the length of the residence permit in cases where the individual was unemployed. Third, it extended to six years (from the previous five) the length of stay before a *Carta di Soggiorno* could be requested. Fourth, the expulsion measures for undocumented migrants and overstayers were further strengthened.

By introducing the contract to stay, the law gave employers increased powers in the conduct of their relationships with migrant workers, since they had the power to directly influence not just their working conditions but also their residence conditions. The integration model, consolidated under the Bossi-Fini Law, weakened the position of migrants, both in the labour market and in their employment relationships. In other words, this model encouraged migrants to accept whatever working conditions they were offered, so long as their employment was regular. Migrants were placed in a situation of permanent negotiation in relation to their own status, since they were constantly exposed to the risk, both of not getting or not maintaining the necessary requirements for their residency status (Basso and Perocco, 2003; Mottura and Rinaldini, 2004; Rinaldini, 2004). This is unsurprising since the construction of a functional relationship between integration and flexibility in the labour market was an explicit aim of the promoters of this legislation.

MIGRANTS IN THE ITALIAN LABOUR MARKET

In 2002, there were around 1.5 million migrants with residency permits. The following year (2003) this number increased by around 700,000, and the total number of regular migrants was 2.195 million (Caritas/Migrantes, 2005). At first the increase was interpreted as the effect of the wide mass regularisations already mentioned. This was true but it is interesting to note that the increase in the following year (2004) was not that much lower: the number of migrants with residency permits totalled 2.79 million (Caritas/Migrantes, 2005).

These data bring to mind two considerations:

- Migrant flows into the country, through legal and illegal routes, corresponded to the demands of families as well as industries and services in the different regions. The extent and the composition of this demand differed, dependent on the regions of the country, reflecting both their economic characteristics and the levels and efficiencies of the services they provided;
- As a consequence, the migration policies already discussed (including their internal contradictions) should be evaluated in relation to the social and economic characteristics of the different region of the country.

Then, before passing to the next paragraph, it is worth setting out some characteristics of migrant workers in the Italian socioeconomic context. Nowadays migrants with regular status represent around 12 per cent on the labour force, with an average unemployment rate not that much different from the Italian one (Caritas/Migrantes, 2007). According to the 2001 census, around 49 per cent of migrants with legal permission were working in the services sector; 45 per cent were in the manufacturing sector; and 6 per cent were in the agriculture sector. The Caritas 2005 report pointed to a partial change in this distribution: a decrease in the percentage working in manufacturing but an increase in those working in construction, together with an increase in those in the services sector (in particular, in the care and hotels sectors) and in the agriculture sector. The same survey, however, confirmed a strong concentration of migrant workers in small and medium-sized industries, in what is called the *triangolo occupazionale* (triangle of employment), the area covered by the regions of Lombardy, Veneto and Emilia Romagna (Caritas/Migrantes, 2005). Moreover, the mass regularisation of 2002 confirmed both the presence of a high number of undocumented migrants employed in manufacturing and a strong presence in the domestic care sector. Therefore, in understanding the role played by migrant workers in the Italian socioeconomic context, it is particularly interesting to observe, on one hand, the strong insertion of migrants in manufacturing and, on the other hand, their insertion in the caregiving field to compensate for the crisis in welfare.[1] This is the context in which an important piece of research was carried out during 2001.

THE RESEARCH 'NON SOLO BRACCIA'

The research 'Non solo braccia' (not just an arm) was carried out in 2001. It was constructed and led by Giovanni Mottura in cooperation with the Istituto di Ricerche Economiche e Sociali-Confederzione Generale Italiana del Lavoro (IRES-CGIL) union of Emilia Romagna. The aim was to investigate various issues regarding the migratory experiences and—in general—the social integration of migrants, with particular attention to their working conditions. The hypothesis adopted was that while stable employment is not of itself a guarantor of successful social integration, it is a necessary condition of beginning a real process of such integration. Here we set out a summary of some key aspects of the research, while others are contained within the original report (Mottura, 2002).

The following data tabulated is based on interviews with 1,654 migrant workers living in Emilia Romagna, using a structured questionnaire survey. Emilia Romagna is recognised as a region offering a *rich context for research* (Brusco, 1989). The survey distribution approximately reflects the composition of migrant employment in the region (Table 6.1).

About half of the migrants interviewed worked in engineering factories, followed by employees in the services sector, in agriculture, in the agro-industrial sector and in the construction, timber and building-material sector. Interviewees employed in small and medium-sized companies were prevalent: 28.8 per cent worked in companies of fewer than fifteen employees; 62.8 per cent worked in companies with fewer than 100 employees and only

Table 6.1 Migrant Workers Distribution by Sectors

	Questionnaire Numbers	%
Steelworkers	662	49.1
Building workers	170	10.8
Food-industry workers	206	12.5
Trade and services	271	16.4
Timber and building materials	31	1.9
Agriculture	272	16.4
Total	1,654	100

Ricerca 'Non solo braccia' 2001.

13.9 per cent worked in companies with more than 500 employees. In all, twenty-three different nationalities were represented. The largest national groups were Moroccans, Ghanaians, Tunisians, Senegalese, Nigerians, Pakistanis, Indians and Albanians. When the research was carried out, migrant flows from Eastern Europe were predominantly female. The pull factor for these women was the demand for domestic care work in Italian families. For Albanian migrants the pull factors were the industrial and building labour market. In general, interviewees were young: only 9 per cent was over the age of forty-five. Finally, as regards gender distribution, 1,348 interviewees were male (81.5 per cent) and 306 were female (18.5 per cent).

SCHOOLING AND EMPLOYMENT IN COUNTRY OF ORIGIN

Data on the educational background of the interviewees revealed that they had a high level of educational qualification. Around 10 per cent had a university education or a degree. Thirty-eight per cent were in the highest three educational categories. However, it is important to bear in mind that, in surveying the educational level of migrants, it is necessary to take account of the diversity of the educational systems in different countries. To homogenise as much as possible the survey data, we classified each educational level in country of origin into three levels (standard, intermediate and advanced), and we again interrogated the data on the basis of this new classification. Table 6.2 shows that the 10.7 per cent of interviewees had an advanced education level, 35.1 per cent had an intermediate educational level and the 54.4 per cent had a standard educational level.

These results confirm both previous and more recent studies about the educational background of migrants. In every study carried out on this

Table 6.2 Educational Level

	No.	%
Standard	884	53.8
Intermediate	581	35.4
Advanced	177	10.8
Total	1,642	100

Ricerca 'Non solo braccia' 2001.

subject their educational level has always been high, and, above all, no relationship has been established between what jobs they are doing and their educational level.[2] Another interesting aspect regarding the interviewees' characteristics is that of their previous employment in country of origin, only 33.1 per cent of the interviewees were unemployed before migration, while 53.5 per cent were employed; 16.5 per cent were students; and 7 per cent had not been looking for work. The data confirm that it is not necessarily the most impoverished who migrate. In particular, among those who had been working prior to migration, around 75 per cent were employed and around 23 per cent were self-employed. Table 6.3 shows that in the first group most were employed in the services sector, followed by those employed in the industrial sector.

In the second group, as Table 6.4 shows, most of the interviewees were self-employed working in the trade sector, followed by the services sector.

However, it is necessary to be careful about the interpretation of the data since it was not easy to break down the characteristics of their jobs in their countries of origin, and the notion of self-employment and of the services

Table 6.3 Migrant Workers Sector in Country of Origin

		No.	%
Employed	Agriculture	98	14.6
	Industry	183	27.3
	Trade	121	18.1
	Other service	215	32.1
	Construction	53	7.9
	Total	670	100

Ricerca 'Non solo braccia' 2001.

Table 6.4 Self-Employed Migrant Workers by Sector in Country of Origin

		No.	%
Self-employed	Agriculture	50	23.1
	Industry	13	6.0
	Trade	86	39.8
	Other services	56	25.9
	Construction	11	5.1
	Total	216	100

Ricerca 'Non solo braccia' 2001

sector may have a multiplicity of meanings. In any case, it is interesting to observe that in cross tabulating the data regarding previous and present employment, it revealed a tendency to look for a job in the same field. However, there does not seem to be any relationship between those who found a job in the same field and the time spent in looking for it. The fact that there is no such relationship, as well as the fact that there is none between educational level and employment obtained in Italy, leads to the conclusion that Italian migration policies have not created an effective mechanism to allow the movement between the demand-and-supply characteristics of the migrant labour force.

INITIAL PERIOD IN ITALY AND FIRST JOBS

Almost all the interviewees (about 93 per cent) had been in Italy for more than two years by the time of the interview. About 30 per cent had arrived prior to 1990, 31 per cent between 1990 and 1995, 31 per cent between 1995 and 1999, while just 7 per cent had arrived less than two years earlier. Depending on when they had arrived in Italy, interviewees might have been able to take the opportunity to regularise their residence on one of the mass regularisations (see earlier). This led to the conclusion that even in the case of those interviewees whose status was regular at the time of the interview, they were likely to have spent some time as an irregular worker while waiting to regularise their position. Data collected concerning their previous employment confirmed this. It was apparent that unauthorised migrants did work. In fact, 80.2 per cent of the interviewees stated that they had always been employed. Table 6.5 shows, the frequency of irregular employment (the only kind of work possible without a residence permit). Indeed, 48.1

Table 6.5 Type of Contract in Previous
Employment in Italy

	No.	%
Regular employment	647	51.9
Irregular employment	600	48.1
Total	1,247	100

Ricerca 'Non solo braccia' 2002.

per cent had been in irregular employment as against 51.9 per cent in regular employment.

Moreover, Table 6.6 shows a further interesting aspect. Cross-tabulating the data related to date of arrival in Italy with those related to the previous employment relationship shows that the more recent the arrival in Italy, the more likely it is that migrants will be in irregular employment.

In other words, by consolidating the data on entry and on the integration of migrants (see earlier) it does not seem that irregular employment has contracted but rather that it has increased. In addition, by cross-tabulating the data related to educational level and that related to the kind of employment relationship (regular and irregular) in migrants' past employment (Table 6.7) it is clear that there is no correlation between the presence of irregular employment and differences in educational achievements.

In other words, it seems that in their first period of residence in Italy (as measured in years), high educational levels did not help avoid the worst working conditions. As regards the working activities previously carried out by interviewees, it emerges that although they all were working, 8.9 per cent had an experience of self-employment in Italy. In this group, 55 per cent had worked in the trade sector and 30.6 per cent had worked in the services sector. These data give an idea of the importance of employment in

Table 6.6 Migrant Workers by Regularity of Employment Contract in Previous Work and Years in Italy

	Years in Italy			
	Less Than 2 Years	Between 2 and 5 Years	Between 6 and 10 Years	More Than 10 Years
Regular employment	29.5%	41.4%	49.4%	66.7%
Irregular employment	70.5%	58.6%	50.6%	33.3%
Total	100%	100%	100%	100%

Ricerca 'Non solo braccia' 2002.

Table 6.7 Migrant Workers by Type of Contract in Previous Employment and Educational Level

	Standard	*Intermediate*	*Advanced*
Regular employment	49.6%	53.8%	55%
Irregular employment	50.4%	46.2%	45%
Total	100%	100%	100%

Ricerca 'Non solo braccia' 2002.

street trading in the migratory processes in Italy, a phenomenon extremely widespread especially during the 80s and 90s. Many studies about migrant street sellers (referred to as *vu' cumprà*) have pointed out how little freedom they had in their employment and how transitory this activity was, and that their aim was always to find different work. As far as self-employment in the services sector is concerned, it emerged that these are mainly 'microservice' activities somewhere between self-employment and 'invented' employment, between regularity and irregularity, for example, like self-employment in car-wash pools or petrol stations, or as babysitters or pizza deliverers, and so forth. The remaining 91.1 per cent said that they had been employees. Those sectors in which the interviewees have been employed in the past were mainly: manufacturing 36.3 per cent, agriculture 36.1 per cent and construction 23.5 per cent. However, it is likely that employment in the agriculture and construction sectors was temporary or seasonal, often irregular. What is interesting is that cross-tabulating the data on current employment with those on previous employment in Italy, the relationship is weak. Moreover, the survey data point to the employment experiences of interviewees being marked by a strong territorial and intersectorial mobility. Such mobility is certainly influenced by two types of strategies adopted at different moments by migrants: the first is to regularise their presence and, as a consequence, immediately to work; the second is trying to improve working conditions, to create some stability and, as a consequence, to search out for a better type of employment. During their first period of residence in Italy these two strategies operate in unison, but the first is definitely much more important, especially if residence is still irregular. Once the residency permit is obtained, the balance between the two strategies changes and the second becomes prevalent. However, both strategies, for different reasons, create a certain employment mobility. In other words, there is a strong employment and intersectorial mobility that often is related to a strong territorial mobility. The fact that they cannot continue in their own occupation (and the consequences of this on their professional identity is to be the subject of further research) results in the adoption of personal strategies by migrants. Nevertheless, these strategies are influenced by the social context and, in particular, by the characteristics of the labour market and by a regulatory framework in which migrants are embedded. Indeed, in the first phase of migration, during which the first strategy is prevalent, this is often distinguished by irregularity, together with employment and territorial mobility in Southern Italy (where irregular employment is very widespread). The second phase of migrants' pathways, when the second strategy is more prevalent, begins when the residence permit has been obtained and when employment and territorial mobility occur in the northern-central regions of Italy (where the context of production offers greater opportunities of regular employment).[3]

CURRENT WORKING CONDITIONS

Table 6.8 shows extremely interesting data on how they obtained their current employment. The most common job search methods were through migrant networks (34.2 per cent), followed by direct approaches to employers (28.7 per cent) and, finally, by personal relationships with Italians (10.9 per cent). If we take account of all who had found their present employment through informal methods (networks, personal contacts and friendships, associations, etc.), the percentage increases to 84.6 per cent. Only the 15.4 per cent had found their current employment through a state employment office or through an employment agency.

Table 6.8 Job Search Methods for Current Employment

Through:	*No.*	*%*
Employment office	146	8.9
Employment agency	76	4.6
Voluntary or aid association	49	3.0
Trade union	68	4.1
Presenting myself to the employer	474	28.8
Answering an ad of the company	59	3.6
Italian friends	179	10.9
Migrant friends	564	34.2
Training course	32	1.9
Total	1,647	100

Ricerca 'Non solo braccia' 2001.

Comparing job-seeking methods for current jobs, it is clear that there are differences by sector, but it is less important to relate informal contacts with specific sectors. The use of *social capital* as a way of seeking employment, and, in particular, the use of *migrant networks* make it important to take account of community. Earlier and recent studies have pointed out that companies use migrant networks to hire workers in order to make recruitment more flexible. Recruitment is carried out through networks of existing migrant employees, mainly based on the common geocultural origins. This method of meeting labour-force supply and demand is a low cost one for employers. However, this informal social dynamic contains many elements of ambiguity. An increase in this kind of recruitment method results in both an inclusion and exclusion processes, since it is often supported by a prejudicial link between specific nationalities and occupations. This means that some national groups are privileged by these social dynamics, compared to others that are excluded, even if those in the privileged group only have access to the labour market through predetermined paths (Ambrosini, 2001). In other words, these dynamics can offer a certain level of flexibility to employers in their recruitment practices, but could consequently promote segregated migrant workforces. Other survey data reveal the type of employment contract interviewees had on starting their current job. As Table 6.9 shows, for 45.5 per cent of respondents, their first employment contract was for a fixed-term contract; for 10.2 per cent it was a seasonal contract; for 8.7 per cent an apprenticeship contract; for 2 per cent is was part time; for 2 percent it was informal off the books. It was only 30.8 per cent who had been given an open-ended contract.

Table 6.9 Type of Contract on Recruitment to Current Post

	Frequency	%
Open-ended contract	505	30.8
Fixed term contract	747	45.5
Part-time contract	32	2.0
Apprenticeship	143	8.7
Seasonal contract	167	10.2
Off-the-books/irreg.	32	2.0
Other	15	0.9
Total	1,641	100

Ricerca 'Non solo braccia' 2001.

Table 6.10 Educational Qualifications on Recruitment

	No	%
Unskilled worker	1,433	87.9
Skilled worker	158	9.7
Clerk	25	1.5
Technician	8	0.5
Middle ranking manager	4	0.2
Senior manager	2	0.1
Total	1,630	100

Ricerca 'Non solo braccia' 2001.

Precarious employment is usually defined in relation to at least one of the following: (a) a time-related aspect (duration/continuity of employment prospects); (b) a social aspect (social rights and protection); (c) an economic aspect (security of income); and (d) a 'working conditions' element (ESOPE, 2004).* If we assume this typology of precarious employment, it is legitimate to state that at least the 69.2 per cent of interviewees had a precarious employment contract when they started their current job, although the level of precariousness might differ. On one hand, these percentages are deeply influenced by the sector in which the company is sited and also by company size (indeed, it would be interesting to study this matter in depth, taking account of the organisation of work as well as the industrial relations dimension in each sector). On the other hand, it is also the case that proportions are not influenced by educational attainment or age. In relation to education (Table 6.10), 87.9 per cent had been hired as unskilled workers and only 9.7 as a skilled workers.

Again, educational attainment and age were not a factor. Among all interviewees, only 37.9 per cent had bettered their contractual position and only 22.4 per cent their qualifications. In addition, among those interviewees whose length of service in their current employment was more than six years and who had seen an improvement in their contracts, only 41 per cent had also seen an improvement in their qualifications, while the 59 per cent were still waiting for this (Table 6.11).

On the basis of all of the data and our analysis, what emerges is a framework which is basically static with regard to qualifications obtained, linked to an increase and persistence of precarious working conditions.

---

*The ESOPE project was a collaborative study on precarious employment consisting of six research institutes.

Table 6.11 Improvement of Qualifications and of Contract (Over Six Years' Length of Service)

		Qualification Improvement					
		No		Yes		Total	
		Number	%	Number	%	Number	%
Empl. Contr. Improv.	No	82	58	59	41.9	141	100
	Yes	62	59	43	41	105	100

Ricerca 'Non solo braccia' 2001.

CONCLUSION

It is useful to recall that the context (not just in Italy) for migration is characterized by a deep transformation of social processes: these changes cover the regimes of work contracts; the financialization of the world economy; the marginalization or even pure alienation of a consistent number of citizens; the impoverishment of social classes; a crisis in the models of industrial relations; and increasing precariousness in the labour market. It seems that the social contract and the centrality of work, the basis for a balanced industrial society, are today in a deep crisis (Gallino, 2004). What is clear is that, in such a context, flexibility and mobility, typical aspects of migrant workers' experience in industrial society, can no longer be considered as a distinct mark of the migratory condition. It is possible to consider that the segmentation of the labour market is still the key tool to understand the fact that local employees are (at least in some sectors) still privileged (Boening, 1967). Simultaneously, it is impossible to not consider that nowadays this segmentation is more and more complex and not based on the dual notion of *guarantee/nonguarantee*. The increasing precariousness and fragmentation of work are the two main aspects of the present global and local processes. Of course, international migration flows are part of these processes, but national (and supranational) migration policies are part of them as well.

In the previous pages, we underlined many factors that denote precariousness, given its changeable forms and aspects, as a permanent factor in every phase of migrants' experience.[4] This does not mean that migrants are destined to be embedded in marginal sectors of the labour market. Nowadays, the hypothesis that migrants' employment is limited to those sectors that Italians abandoned has been challenged by the development itself of the migration processes in Italy over the last thirty years. However, it has demonstrated that the link between migration and precariousness has assumed various forms. In fact, the paradox is that the declared desire to put an end to the precariousness of migrants (in the past considered the result of their undocumented status) has produced, through laws, an indissoluble link between regular residence and the employment relationship and as a consequence a prolonged precariousness. Regular migrant workers are constantly exposed to the inevitable risk of the failure of their projects. Paradoxically, the power of such a threat may be stronger the more integrated migrants are in their local contexts (family reunion, schools for children, housing loan, social relationships, etc.). At the same time the legal restrictions mean an increase in undocumented migrants, overstayers and migrants without the opportunity of renewing their residency permits in cases of prolonged unemployment (which always means simply seeking irregular employment). The final result has been the creation—and the constant renewal—of a *wide pool* making up a heterogeneous and fragmented migrant labour force, subject to the implicit threat of a waver of residency, despite being employed in occupations requiring professional skills, specific knowledge and experience.[5]

This conclusion clearly evokes the Marxian theory of the *reserve army of labour* or *relative overpopulation* (Marx, 1967). Actually this interpretation is not new in the Italian literature on migration. This category was used frequently to analyse Italian internal migration to Northern Italy and North European countries during the 50s and 60s. During the 1970s it was widely adopted to analyse changes of the Italian labour market (Meldolesi, 1972; Daneo, 1971; Vinci, 1975; Mottura and Pugliese, 1975; Donolo, 1972). Today this theory seems again to be useful in understanding the present migratory processes in Italy. In particular, it is useful to highlight some aspects of the theory in order to better understand the role of the migrant workforce within the dynamics and stability of the Italian labour market. According to Marx, there are thousands of particular shades of *relative overpopulation* (it is necessary to remember that the Marxian meaning of the term *overpopulation* radically diverges from the Malthusian meaning of the term). The author underlines only three general forms of 'relative overpopulation' containing other more particular forms of it. The analysis aims to highlight the fact that the existence of the *relative overpopulation* (and the use of it) is understandable only if it is related to the needs of the enhancement of capital. Then the analysis aims to demonstrate that the study of any phase of capitalistic development may not ignore the extent, the form and the characteristics of *relative overpopulation*. According to the author, just taking care of these dynamics allows us better to understand both the form of the capital at every phase of accumulation process and the balance (the relationship between the classes) characterizing the social and institutional asset of the system at the given time. If we adopt this point of view, it seems clear that the meaning of the *relative overpopulation* category is not reducible to a simple synonym for *unemployment* (a common misunderstanding among economists). In fact, according to Marx, unemployed people are only a component (not prevalent) of the *relative overpopulation*, except in critical phases. Moreover, explicit (in the author's text) is the impossibility of identifying the category of *relative overpopulation* with the category of *marginal mass*, a widely used definition in several sociological studies (Nun, 1970). It then seems possible to see that nowadays migrant workers in Italy are a component of the *relative overpopulation*, apart from the legality of their juridical status and the regularity of their employment. This means that the factors that allow us to explain the presence and role of migrants, together with the restrictive feature of migrant policies (that could appear contradictory given that the shortage of local labour is generally recognised), is its function in encouraging the 'babelization' of the *relative overpopulation* composition, during a phase characterized by transformations, producing a widespread social and working insecurity. In fact, undoubtedly most local workers are also embedded in processes of growing social and work insecurity. Most local workers do not differ greatly from migrant workers, insofar as their employment contracts and working conditions are concerned. However, to paraphrase a Marxist expression, any possibility of mutual recognition between migrants and local workers seems destined to be frustrated or at least to be made difficult by the *capitalistic use of cultural and national differences*.

NOTES

1. The strong increase of demand for domestic caregiving labour (for disabled and elderly persons) is interpreted as a clear sign of welfare system crisis (Mottura, 2006; Pugliese and Mottura 2005).
2. This lack of relationship between the educational attainments of migrants and their level of insertion into the labour market has always occurred in the course of migration processes in Italy. In fact, this issue was first pointed out in research carried out in the 80s and it has subsequently always been confirmed. Moreover, research carried out in 2004, by E. Reyneri, highlighted the risk that paradoxically, within the Italian labour market, the high educational levels of migrants could constitute a block on their process of stabilization (Reyneri, 2004).
3. In other ways the role of migrant workers in the labour market depends on the geographical region. It is common knowledge that in Italy the insertion of migrants into the labour market and their social integration depend on the social and economic characteristics of the different areas: northern/central/southern Italy, big cities/small cities, industrial/agriculture context, etc. (Pugliese, 1996; Mottura, 1992).
4. In a recent study on different form of precariousness, the concept used is one of *fragility* (Carchedi, Mottura, and Pugliese, 2003).
5. An example could be care workers employed in families as well as nurses often employed under a status and salaries that are definitely inadequate for the required tasks (B. Ehrenreich, and Russel, 2004); another strong example would be that of skilled bricklayers and other workers getting unskilled wages.

7 Citizenship and the Disciplining of (Im)migrant Workers in the United States

Nandita Sharma

> Instead of bending all its subjects into a single uniform mass it [disciplinary power] separates, analyses, differentiates, carries its procedures of decomposition to the point of necessary and sufficient single units (Michel Foucault).

The experiences of those who migrate to the U.S. vary dramatically. Many factors contribute to these, but none of them can be said to be located solely within the individual. Rather, these factors are structural in character. Where one migrates from, how that location is situated in the broader geopolitical concerns of the U.S., as well as how that location fits into the broader global capitalist economy; how one's formal educational qualifications are judged within the U.S.; how one is racialized and gendered; and, last but by no means least, how one is classified and positioned by the state in its hierarchy of differential statuses (citizen, permanent resident, refugee, temporary 'guestworker' or 'illegal') all have enormous consequences for the labour market (and other life) experiences of (im)migrants.[1]

Of course, recognition of these structural factors does not correspond to an easy correlation between individual and group experiences of migration to the U.S. Individuals within a group of similarly situated people may very well manoeuvre a beneficial position for themselves in the U.S., in ways that most group members cannot (often through some confluence of having networks of support and sheer luck). However, the telling of individual successes in negotiating life in late capitalist USA does not translate into a systematic analysis of the barriers faced by recent (im)migrants in securing and being secure in relatively well-paid employment that provides the means for a healthful life in the country.

The reasons for the precarious labour market experience of the vast majority of (im)migrants are to be found in the kinds of policy directives taken by the U.S. government and its immigration officials, in the structural organization of the labour market, in the demands made by employers and in the white nationalist culture of the U.S. that racializes normative values of citizenship which work to continuously discriminate against those (im) migrants who do not fit the limited understanding of national American subjectivity. Indeed, it is crucial that we examine (im)migration as a cultural phenomenon, as well as a political and economic one.

It is therefore necessary to examine who benefits from current U.S. immigration policies and who does not. It is also useful to examine the economic, political and cultural context into which (im)migrants enter once moving to the U.S. In this chapter, I will do this by examining the ongoing saliency of national border regimes in the U.S. I will show that U.S. border regimes that regulate—and regularly demonise—the entry of (im)migrants are a key part of U.S. state neoliberal policies that help to reorganize the U.S. labour market (as well as workers across the world) to create a more 'flexible' and globally competitive labour force. In this way, we can see that ever-more-restrictive (im)migration policies and militarised borders are not a contradiction to the expansive global policies concerning cross-border trade but are, in fact, their complement.

To better understand the overwhelming legitimacy given to policies that continuously create precarious (im)migrant workers in the U.S., I examine how the ideology of the U.S. as a 'land of immigrants' and as a hospitable, welcoming place that provides sanctuary to the world's 'huddled masses' and 'wretched refuse' operates to mask how U.S. immigration policy (historically and today) is *central* to the ongoing supply and reorganization of the supply of labour available to employers in the U.S., in ways that are largely detrimental to (im)migrants. The process of racializing both American national subjectivity and that of those deemed to be 'enemies' plays a key role in this restructuring. This is because the constant construction of racialized criteria for membership and nonmembership in the American 'nation' serves both as a mechanism of social control and a key element in the ongoing legitimacy of the American national project. Indeed, the politics of anti-immigration in the U.S. have always been a part of how the American polity is defined, and while the targets of such political hatreds have shifted (as have, in some cases, those who espouse such politics), they remain central to our understanding of the labour market position of new (im)migrants.

The chapter provides, first, a broad sketch of the current labour market position of various (im)migrant workers in the U.S., followed by a historical examination of legislative developments that have led to the positioning of the vast majority of contemporary (im)migrants as cheapened and weakened persons labouring in the fields, factories, fast-food restaurants, hospitals and care homes, schools, convenience stores, and even, rarely, in the board rooms of the U.S. By employing the theoretical understanding of Gilles Deleuze and Felix Guattari that forms of exclusion are better understood as forms of 'differential inclusion', where the exercise of state disciplinary power works to separate, differentiate and individualize the social processes that subordinate most (im)migrant workers, I will also discuss how the dominant rhetoric of 'border order' in the U.S. works ideologically to legitimate what can only be called a form of 'global apartheid' operating in the U.S. labour market, an apartheid that creates hierarchical legal positions for various groups of people according to their national status in the country. By way of a conclusion, I will examine recent efforts at organizing for progressive change by (im)migrants and their supporters and the attempts to shift and even open up the discourse on (im)migrant rights to incorporate all subordinated workers in the U.S., as well as globally. The emergence of significant trends within the (im)migrant rights movement to turn it into part of the global movement for social justice provides us with a way of linking U.S. immigration policy to not only other U.S. policies (military, foreign, trade, etc.) that work to displace people but to expand our understanding of how national immigration policies can only rightly be understood within the global geopolitical, economic and cultural contexts.

THE POSITION OF NEW (IM)MIGRANTS IN THE U.S. LABOUR MARKET

Looking at the characteristics of that portion of the population in the U.S. that was not born in the country is a fairly good indicator of how new (im)migrants are faring. Rakesh Kochhar (2007b: ii), in analysing U.S. Census Bureau numbers from 1995 to 2005, notes that 15 per cent of the U.S labour force is made up of new (im)migrants ('foreign-born', in the language of Kochhar's report). This is an increase from 7 per cent in 1980 and is therefore a *doubling* of new (im)migrant workers in the past quarter century (ibid.). Indeed, from 1990 to 2000, there was a 61 per cent increase in the (im)migrant population in the U.S. This has had a major impact on the U.S. labour market since new (im)migrant workers now '. . . account for the majority of new workers in the economy' (Kochhar, 2007b:1). Along with an increase in the overall number of (im)migrant workers, these workers have also come to live and work in areas of the country that previously had relatively few (im)migrants. While U.S. states with an already high proportion of (im)migrant workers, like California, have seen a substantial growth—44 per cent or 2.5 million new (im)migrant workers in 2005—Kochhar (2006: 2) reports that '[i]n all, in 18 states the foreign-born population 16 and older grew by at least 100% between 1990 and 2000'. Both of these trends point to an increased dependency by U.S.-based employers on new (im)migrant workers.

Such workers are much more likely than U.S.-born workers to work in low-wage jobs (ibid.). About half of new (im)migrant workers were either in the low-wage group or in the low-middle group in 2005. The probability of being a low-wage worker was highest for those who had arrived in the U.S. after 2000: 40 per cent of all workers who had arrived in the 2000–2005 period were in the low-middle earning group (Kochhar, 2007b: 7). Overall, only 11 per cent of most recent arrivals were high-wage workers in 2005 (ibid.).

Most recent (im)migrants are nonwhites: the main 'sending' regions and/or countries of (im)migrants to the U.S. in 2005 were: Mexico (31 per cent), South and East Asia (23 per cent), the Caribbean, Central and South America (23 per cent) and the Middle East (less than 4 per cent). 'Others' not categorized were almost 20 per cent (Pew Hispanic Center, 2006b). Given the systemic character of racism within the U.S., it is unsurprising that there are significant disparities amongst not only those born in the U.S. but also in the experiences of new (im)migrant workers.

In 2005, 75 per cent of U.S.-born white workers worked as part of the high-wage workforce (Kochhar, 2007b: 24). Despite the rhetoric of (im)migration causing a decline in their life chances, the employment growth of U.S.-born whites was primarily in this high-wage group (Kochhar, 2007b: 1). Indeed, '. . . they added more workers to the high-wage workforce than might have been expected, based on demographic trends alone', thereby strengthening the 'wages of whiteness' (Ibid.). Like their white counterparts, U.S.-born black workers were unaffected by the entry of new (im)migrants into the workforce. However, very much unlike whites, half of these black workers laboured in the lowest wage groups, while about 30 per cent of U.S.-born black workers were in the two highest wage groups (Kochhar, 2007b: 24).

In the U.S., Hispanic workers are the most likely to work as low-wage workers. Approximately 66 per cent were either in the low-wage group or the low-middle wage group (Kochhar, 2007b: 6). Very few were able to secure high-waged work—only 6 per cent in 2005 (Kochhar, 2007b: 7). For people arriving from Mexico, the likelihood was even less—only four per cent (ibid.). There have been certain improvements in the situation of newly arrived Hispanic (im)migrant workers. Kochhar (2007b: i) reports that '[t]he proportion of foreign-born Latino workers in the lowest quintile of the wage distribution decreased to 36% from 42%, while many workers moved into the middle quintiles'. This is largely attributable to a combination of individual and structural factors. Most recent Hispanic (im)migrants are generally older, have higher levels of formal education and, unlike previous cohorts, are more likely to be employed in construction than in agriculture (Kochhar, 2007b: i).[2] The meltdown in the credit market in the U.S., with its consequent decline in new housing starts, bodes ill for the continuation of this trend, however. Moreover, Kochhar (ibid.) further notes that '[e]ven though the share of Latino workers at the low end decreased, in absolute numbers this population grew by 1.2 million between 1995 and 2005'. The result, then, is that overall, recent Hispanic (im)migrants added more low-wage than high-wage workers. This picture does not look vastly different for U.S.-born Hispanic workers, almost half of whom are to be found in either low-wage or low-middle wage groups (Kochhar, 2007b: 22).

Low-wage work exists in all sectors of the U.S. economy. Whether employed in jobs in construction; educational, health and social services; arts, entertainment, recreation, housing and food services; retail trade;

professional, scientific, management, administrative and waste management services; nondurable or durable goods manufacturing; finance, insurance, real estate and rental and leasing; or agriculture, most immigrants from traditionally nonwhite locations had low incomes. The median personal earnings of (im)migrants by country or region of birth in 2005 were: Mexico, $17,000; Central America, $19,000; South and East Asia, $30,000; Caribbean, $24,000; South America, $24,000; Middle East, $30,000 (Pew Hispanic Center, 2006d—Table 25. Median personal earnings by region of birth: 2005).

To flesh this out further and show the differential effects of processes of racialization in the U.S., let us look at the personal earnings of new (im)migrants by their country or region of birth in 2005. Almost 57 per cent of (im)migrants from Mexico earned less than $20,000 and 94 per cent earned less than $49,999. Other Hispanics again proved to have the highest numbers of (im)migrants with low incomes: just over 52 per cent of people from Central America and 40 per cent from South America earned less than $20,000 while just over 92 per cent from Central America and almost 84 per cent from South America earned less than $49,999. Almost 40 per cent of (im)migrants from the Caribbean earned less than $20,000, while approximately 83 per cent earned less than $49,999. Almost 34 per cent of people (im)migrating from the Middle East earned less than $20,000, while almost 67 per cent earned less than $49,999. (Im)migrants from South and East Asia were the most likely to earn more than $50,000 but these, again, were in the minority of (im)migrants from Asia. In contrast to the myth of Asian (im)migrants being a 'model minority', about 31 per cent earned less than $20,000, while about 68 per cent earned less than $49,999 (Pew Hispanic Center, 2006d—Table 24. Personal earnings by region of birth: 2005).

The labour market experiences of nonwhite (im)migrant workers to the U.S., then, most closely mirror those of nonwhite U.S.-born workers. The number of South and East Asians in the high-wage bracket are closest to U.S.-born black workers (again, the group most likely to have high-wage earners) while (im)migrant workers from Mexico are significantly underrepresented, and U.S.-born white workers are significantly overrepresented in this category. Unsurprisingly, then, poverty rates amongst nonwhite (im) migrants are relatively high: They are the highest for people (im)migrating from Mexico (25.5 per cent); almost 18 per cent for those from the Caribbean; 17 per cent for those from Central America; almost 17 per cent of people from the Middle East; almost 12 per cent for those from South America; and 12 per cent for those from South and East Asia (Pew Hispanic Center, 2006d—Table 30).

On average, there is a substantial gap in the poverty rates for U.S.-born versus (im)migrants: those born in the U.S. have a poverty rate of about 13 per cent while that of (im)migrants is slightly above 17 per cent (Pew Hispanic Center, 2006d, 'Foreign Born at Mid-Decade'). Along with how one is racialized, then, citizenship matters. Thus, while some researchers (Borjas, 2003; Sum, Harrington and Khatiwada, 2006) argue that recent immigration has proven detrimental to U.S.-born workers, others (Card, 2001; Ottaviano and Peri, 2006; Kochhar, 2006) have shown that there has either been no effect or that U.S.-born workers have, on average, benefited from immigration (Council of Economic Advisers, 2007, as quoted in Kochhar, 2007). At the very least, there is strong evidence to show that the rapid increases in (im)migrants over the last decade '. . . are not associated with negative effects', either in the economic boom of the 1990s or in the period of recession and slow recovery that followed in 2000 (Kochhar, 2006).

Moreover, contrary to the rhetoric of (im)migrant workers being particularly detrimental to the employment opportunities and wages of disadvantaged U.S.-born workers, Kochhar (2006: i), in his extensive study, found that '[t]he size of the foreign-born workforce is also unrelated to the employment prospects for native workers. The relative youth and low levels of education among foreign workers appear to have no bearing on the employment outcomes of native workers of similar schooling and age'. Likewise, the share of (im)migrant workers is not found to be related to the employment rate of U.S.-born workers in the slowdown and recovery years, respectively, of 2000 and 2004 (Kochhar, 2006: 2).

In short, citizenship, with its attendant meanings of national belonging, profoundly shapes the experiences of people who fall on either side of the citizen/noncitizen divide.[3] For both sets of people, citizenship operates as a technique of labour market organizing. In the U.S., (im)migrants have higher poverty rates and, on average, have much lower average wages and many less employment-related securities than citizen workers. One recent study found that only one-third of new employed (im)migrants reported having a health-insurance plan at work while less than one in five were covered by a pension plan at work. This contrasts with nearly one-half of the U.S.-born workforce (Sum, Harrington and Khatiwada, 2006: 9). Indeed, the (im)migrant group with the highest wage levels is at par with one of the more disadvantaged groups of U.S.-born workers—U.S. blacks.

We need to pay close attention to these effects of citizenship rather than naturalize them as a sort of evolutionary process that sacrifices new (im)migrants to marginal positions within the labour market only to be surpassed by their second- or third-generation offspring. Indeed, studies done elsewhere in the Global North show, for instance, that in Canada, second generation, nonwhite Canadians have even lower average wages than their (im)migrant parents, despite the Canadian schooling, official language skills and cultural experience. This form of discrimination is racialized, but it is a racism structured through nationalist modes of imagination *and* the power of the U.S. national state to enforce ideas of national membership (and, therefore, nonmembership). I turn now to examine U.S. immigration policy, past and present, to look at how it structures these discriminations and, just as importantly, how these are legitimated in the U.S. polity.

ILLEGALISED (IM)MIGRANT WORKERS

Illegalised (im)migrants to the U.S. comprise the largest group of (im)migrant workers. The Pew Hispanic Center (2006: 2—fact sheet), utilizing data from the March 2005 Current Population Survey, conducted jointly by the U.S. Bureau of Labor Statistics and the Census Bureau, estimated that in 2005, there were about 11.5 to 12 million illegalised (im)migrants in the U.S. This figure dwarfs those for either legalized permanent residents, who numbered a little more than 705,000 persons in 2003, or temporary H-visa (or 'guest workers'), who numbered almost 360,000 in that same year. Even more striking is the comparison between permanent residents who were specifically admitted with permanent residency rights on employment-based criteria, those admitted either as 'guest workers' or who entered, lived and/or worked as illegalised persons. In 2003, only about 82,000 persons were granted employment-based 'green cards' (permanent residency cards). In other words, only 19 per cent of (im)migrants in 2003 were admitted and permitted to work as immigrant workers who held the rights of permanent residency. In contrast, 81 per cent were admitted and tied to their employers through the strictures of the three kinds of H-visa programs.

If we compare this with the number of illegalised persons living in the U.S. (the broad number of which is admittedly larger than the number of those working in the labour market), it becomes clear that the number of illegalised workers dwarfs—12 million illegalised (im)migrants compared to the less than 500,000 legalized (im)migrants in 2003, given either work-related green cards (82,000) or H visas (360,000). Consequently, the overwhelming majority of (im)migrant workers in the U.S. work under conditions where they are denied virtually all of the labour-force rights, protections and entitlements of permanent residents or even the limited amount of practical or legal rights and protections (minus entitlements) of H-visa workers.

U.S. employers' desire to illegalise or otherwise make more vulnerable (im)migrant workers in the U.S. is by no means a new phenomenon (Bustamante, Reynolds and Ojeda, 1992). However, their reliance on illegalised (im)migrant workers has both increased as well as expanded across most sectors of the U.S. economy. The number of illegalised workers has increased since the economic boom of the late 1990s, particularly in the service and construction industries (Alvarez, *NYT*, December 20, 2006). Indeed, of the estimated 11.5 to 12 million illegalised (im)migrants residing in the U.S. in 2005, it is believed that almost 4.5 million are fairly recent, having come in the period between 2000 and 2005 (Pew Hispanic Center, quoted in Sum, Harrington and Khatwada, 2006). It is currently thought that an estimated one million illegalised (im)migrants continue to come to the U.S. each year, with finding paid employment being the largest reported reason to do so.

Because of the high proportion of illegalised (im)migrants working in the U.S. labour market, we need to recognize the *sources* of their vulnerability, precariousness and cheapness. Illegalised persons face severe discrimination on the basis of their status, a status, it is worth remembering, that is accorded to them by the U.S. state. Indeed, making someone illegal has been a relatively lengthy—and always *social and legal*—process. Beginning with the Immigration Act of 1924 that made it a felony to enter the U.S. without inspection, to the 1954 'Operation Wetback' immigration police program that sanctioned and organized mass deportations of workers from Mexico recruited through the Bracero Program, an agricultural guest worker scheme in operation from 1942–1964 (described as 'legalized slavery' by its own former administrator, Lee G. Williams), to President Reagan's Immigration Reform and Control Act (IRCA) of 1986 that for the first time made it illegal (and punishable) for undocumented migrants to hold a job or for an employer to hire them and increased resources to the Immigration and Naturalization Service (INS, which was absorbed into the Department of Homeland Security in March 2003), to President Clinton's increased use of raids against illegalised persons and his 1996 Illegal Immigration Reform and Immigrant Responsibility Act (IIRIR) that broadened the criteria that require detention and eventual deportation, illegalised workers have been progressively more criminalized and driven into an underground economy, where working conditions and rates of pay are far less than those enjoyed by citizens and permanent residents.

Immigration status, then, is a workplace issue (Bacon, 1999). Illegalised workers work 'scared and hard,' to quote the ex-secretary of labour, F. Ray Marshall (in Behdad, 2005: 32). The 1986 Immigration Reform and Control Act and subsequent increases in its enforcement have contributed significantly to this situation. In the late 1990s, Cristina Vasquez, a regional manager with the Union of Needletrades, Industrial and Textile Employees, maintains that, 'immigration law is a tool of the employers. They're able to use it as a weapon to keep workers unorganised, and the INS has helped them' (in Bacon, 1999). Since the mid-1980s and increasingly throughout the 1990s and into the 2000s, what William Gould, past U.S. National Labor Relations Board chair, understood as the '. . . basic conflict' between workers' rights under the National Labor Relations Act and workplace enforcement of immigration law' has been exacerbated (ibid.).

As the penalties for being discovered as an illegalised person in the U.S. grow and as the state increasingly utilizes the technique of deportation (more than 162,000 people were deported in 2005; see Koch, 2006), labour and other political organizers have remarked on the creation of a chilly climate for organizing (im)migrant workers on labour and other issues of import. This is reflected in a recent Pew Hispanic Center (2007) survey that found that '[j]ust over half of all Hispanic adults in the United States worry that they, a family member or a close friend could be deported'. Anywhere from one in eight to one in four also say that the anti-immigration political environment in the U.S. has hurt them personally. Latino (im)migrants report having greater difficulty finding employment and housing, being less likely to use government services or to leave the U.S. They also report an increase in harassment, centring on their status in the country, as they report a greater likelihood of having to produce legal documents proving their (im)migration status (ibid.).

Such findings are significant because Latinos are the largest minority group in the U.S. (approximately 47 million persons or about 15.5% of the U.S. population; ibid.). They are also the most likely to be illegalised in the U.S.—about a quarter of Latino adults are illegalised (im)migrants (ibid.). What happens to Latinos in the U.S. can arguably be seen as a template for the treatment and experience of other (im)migrant, as well as minority-citizen, groups. What this experience shows is that illegality has a serious effect on the wages that a person receives. Francisco Rivera-Batiz (1999) found that legalized male (im)migrants from Mexico earned an hourly age rate that was almost 42 per cent higher than those of illegalised workers. Similarly, female legalized (im)migrants earned almost 41 per cent more than their illegalised counterparts. While some of this gap could be accounted for by demographic, occupational and/or human capital differences, this did not explain the entire wage differential. Rivera-Batiz found that more than 48 per cent of the gap between male legalized and illegalised workers or 43 per cent of the gap between the equivalent female (im)migrant workers was due to the person's status. Additionally, in studying the effects of the 1986 policy legalization of previously illegalised workers, this same study found significant wage growth for affected workers in the four years following the reforms. As the characteristics of the workers did not change, it is clear that legalization led to higher wages for the same group of workers. Illegalised workers in the U.S., then, have lower wages than those who are legalized *because* of their state-mandated illegal status (also see North and Houstoun, 1976, S-11; Massey, 1987). Status, thus, operates as a significant form of discrimination in the U.S. labour market, a form of discrimination, moreover, that is almost invisible, given the hegemonic status of ideas of citizenship and notions of national state sovereignty.[4]

Illegalising migrations, indeed illegalising the vast majority of migrations, to the U.S. provides employers, in an ever-expanding range of industries and sectors, with a cheapened and weakened group of workers whose militancy is severely hampered by state controls such as workplace raids, detention and deportation. Hence, while illegal migrations are often thought to signify an *absence* of migration controls, it is clear that illegalised migrations are a *part*—a significant part—of the U.S. state's (im)migration program. Why then are there major proposals to shift the status of (some of) those currently working in the U.S. under an illegalised status to 'guest workers'? The answer lies in how already-existing 'guest worker' schemes operate. I turn to an examination of the system of H visas to understand why many employers groups are urging Congress to broaden such programs.

H-VISA SYSTEM OF RECRUITING 'GUEST WORKERS'

The H-Visa system of recruiting people as 'guest workers' exemplifies the U.S. adoption of 'managed-migration' schemes. As mentioned previously, most people legally (im)migrating to the U.S. now arrive not as immigrants (those with U.S. 'green cards' or permanent resident status) but as nonimmigrant 'guest workers' with one of three kinds of H visas—H-1B, H-2A or H-2B visas. The contemporary era of the H-visa system began in 1986, with the passing of the Immigration Reform and Control Act (IRCA). By law, a guest worker is the 'foreign worker' who upon arrival must have with her or him an official H visa from the American state. This H visa constitutes a labour contract that ties a person to a specified employer and stipulates her/his occupation, residence, length and terms of employment in the U.S. Most H-visa workers must exit the country immediately after their labour contract expires. Written permission from immigration officials is required to alter any of the conditions of work, including changing the employer or occupation, working for additional employers, and so on. If any of the terms are changed without official permission, guest workers are subject to immediate deportation. Of course, deportation results in job loss.

The three types of H-category visas correspond to the occupational sector the worker is employed in. The H-1B visa, created in its present form in 1990, recruits migrants for a variety of white-collar jobs (INS, 1991: 3).[5] Currently admitting about 65,000 people a year, the high mark was reached in the mid-1990s when about 195,000 people were admitted annually. The H-1B visa is distinct in several ways from other H visas. Its holders are permitted to legally work for much longer periods of time—an initial three years, with another three-year renewal allowed. The H-1B visa is also the only one that allows workers to apply for a shift in status from temporary to permanent resident from either within the country or abroad. Moreover, unlike other H-visa holders, H-1B workers are allowed to bring their spouses and children, although they can only legally work if they hold their own work visa. Children do, however, have the right to attend school in the U.S. The H-2A, created in 1986, recruits migrants for temporary or seasonal agricultural work (INS, 1991: 4). Currently, about 84,000 enter under these visas (Employment Policy Foundation, 2002). The H-2B visa, created in 1991, recruits migrants working for nonagricultural, low-skilled jobs which are claimed by employers to be '. . . seasonal, intermittent, to meet a peak load need, or for a one-time occurrence' (INS, 1991: 5). Last year, approximately 65,000 people worked under these visas. H-2B visa workers can only stay in the U.S. for a maximum of 364 days, with their average stay being approximately eight or nine months.

These are clearly important differences between the three kinds of H visas. However, it is equally important to recognize that all H-visa holders, regardless of their occupation, are entirely dependent on their employment status, and therefore on employers, for their legal standing in the U.S. More-over, people admitted through the H-visa schemes are denied the freedoms of labour market and spatial mobility available to those existing within the legal designation of citizen or permanent resident. In fact, the H-visa system exists because it is *unconstitutional* for the federal government to restrict the labour market or geographical mobility of citizens or permanent residents. Such restrictions apply *only* for those legally classified as *non*immigrants who by law exist in unfree employment relationships. The H-visa or guest-worker category, therefore, allows the social category of *foreigner* to be fully realized in American law. H-visa workers exist in the U.S. under a very dif-ferent legal regime than do citizens or permanent residents. They can be made to live and work as unfree workers in the U.S. because they have been classified as nonimmigrants or 'guest workers'.

Indeed, the one part of the program that employers find most appealing is that guest workers are legally tied to them. This is recognized by the U.S. General Accounting Office (2000), which states that H-2A (but arguably all) workers in the H-visa program 'are unlikely to complain about worker protection violations, fearing they will lose their jobs or will not be hired in the future'. In the words of one worker, 'What you see, you must remain silent' (Yeoman in Pastor and Alva, 2004: 96).

Alongside producing unfree workers, the H-visa system also provides 'guest workers' to employers concerned with securing a post-Fordist labour force: efficient, flexible and almost wholly lacking in options. Part of the flexibility and competitiveness of guest workers is that they are denied access to many of the things that capitalist lobby groups complain make Ameri-can workers 'too expensive' or 'too inflexible': access to mobility rights and social-welfare programs being two of the most important. Guest workers are ineligible for social benefits, such as unemployment insurance or welfare payments, even though they do pay taxes and have money deducted from their pay cheques for such programs.

Employers benefit in numerous ways from such legally organized employ-ment relations. Case studies on select groups of H-visa workers show us that they are generally paid much less than citizens or permanent residents and are made to work and live under conditions generally seen as 'unattractive' to those with the freedom to move—move locations and move jobs.[6] As Deborah Waller Meyers (2006: 3) puts it, H-visa jobs are '. . . commonly . . . described as dirty, demanding and dangerous'. In the few case studies that have been done on the various H visas, there are countless accounts of abuse, in relationship to numbers of hours worked, wage rates, poor working and living conditions, etcetera. While such abuses are also found for vulnerable workers with citizenship or green-card status, workers on H visas have no legal options to move in the labour market or to move geo-graphically without either being deported or rendered illegal. It is therefore important to note the significant shift in (im)migration status brought about by the H-visa program. In 2003 about 34 per cent of all legal migrants entering the U.S. were denied permanent status and were instead issued a temporary, indentured status to work in the country. However, comparing the number of people issued green cards as a result of being sponsored by their employers (versus being sponsored by a family member) with the numbers admitted as H-visa workers helps us to analyse the significance of H visas to the 'employment' stream of (im)migration and allows us to see that in the less than two decades since its inception, the number of people legally admitted to work in the labour market in the U.S. as immigrants (i.e. permanent residents) has declined both in proportion and in number to those recruited under H visas. So-called 'employment-based' green-card holders only accounted for approximately 82,000 or 12 per cent of all green card holders in 2003.

Given the current policy debates and developments in immigration policy that are working to discursively configure the expansion of current guest-worker programs as a *humanitarian* option to the legalization of people, there is a strong possibility that the U.S. system of guest workers will be further expanded, likely through the creation of a separate temporary work-visa category. President Bush's first major policy statement of the 2004 election year centred on 'matching willing employers with willing workers' by expanding the guest-worker system. 'Willing workers' coming to mean workers who are legally tied to their employers. The U.S. Senate picked up on the president's proposal by passing the Comprehensive Immigration Reform Act (CIRA) on May 25, 2006. In both the president's statements and the CIRA, there is a proposal for an additional 325,000 guest-worker visas annually. Interestingly, the CIRA was seen as the 'kinder, gentler' version of the U.S. House of Representatives Border Protection, Antiterrorism, and Illegal Immigration Control Act (HR 4437) passed on December 16, 2005, that, if implemented, would have turned undocumented migrants into felony criminals by making the simple entrance and residence in the U.S. without state permission a criminal act.

However, when we examine the details of both House and Senate Bills, the expansion of the guest-worker program is directly linked to the discourse of the immigration system being 'out of control' and the need to bring 'order to the border'. Such discourses have seemingly become commonsensical, as evidenced by three polls released in 2006: one, by CBS News and the *New York Times* (2006), reported that 59 per cent of Americans surveyed described the problem of illegal immigration as 'very serious'. Another, conducted by *USA Today*/Gallup (2006), found that 81 per cent characterized illegal migration as 'out of control'. Interestingly, a *Los Angeles Times/* Bloomberg (2006) poll found 58 per cent of the respondents supporting an approach that combined tougher enforcement of immigration laws along with a guest-worker program for illegalised migrants.

In this regard, it is important to note that while being portrayed as world's apart, *both* the current Senate and House bills provide for an increase in border patrols, further criminalize those living in the U.S. without official state permission and lessen the total number of green cards issued. We can say, therefore, that the debate on whether to expand guest-worker programs in the U.S. or not hinges on the legitimacy of denying permanent residency to a growing number of migrants wishing to live and work in the U.S. The 'grand bargain' that took place in the U.S. Senate over the spring of 2006, therefore, is likely to lead to an even weaker, structurally embedded labour force of both illegalised as well as guest workers. As there will be likely be even fewer legal means to enter the U.S. to work, while the processes leading to unemployment and other forms of displacement are not ameliorated, the number of (im)migrants coming to the U.S. as illegalised persons or applying to enter as 'guest workers' is highly likely to grow. This means more and more people living and working within the U.S. are without the same sets of rights and entitlements given (in varying degrees) to permanent residents and citizens. The greatest beneficiaries of this will be employers, who will have greater access to a greater number of workers denied most, if not all, social and labour market protections as well as the state itself, which will collect taxes from both illegalised and 'guest' workers while denying them the right to benefit from state welfare services. The profits and savings in state spending are enormous. The costs of this program will, in contrast, be borne most heavily by subordinated (im)migrant workers and their social networks.

Since the very first immigration practices of the U.S. state, creating conditions of work that are unacceptable, even unconstitutional, for American citizens have been easier to impose on noncitizen women and men who are also seen as racialized, cultural outsiders. Hence, because 'guest workers' are not entering a 'neutral ideological context' (Miles, 1982: 165) when coming to the U.S., H visas and the guest workers that they produce need to be located within the ideological organization of the Otherness of nonwhites in American society.

It is important to note that despite the unconstitutionality of forcing citizens and green-card holders to work in the conditions legally and routinely imposed on all H-visa workers, the dominant discourse portrays them as 'lucky' to be living and working under official constraints. The act of allowing Them into the country as, literally, Our 'guests' is seen as an act of charity extended by Americans to foreign Others—foreign others, not insignificantly, who are seen as part of the nameless, faceless mass of downtrodden 'Third World–looking people' desperate for a place in the U.S. (see Hage, 2000). Within this discourse of Third World luck and American charity, the socially organized conditions for the desperation for those (im)migrant workers denied the rights of permanent residency is soundly ignored, a desperation wholly exorcised from the formation of American citizenship and immigration policies and their effects on the labour market in the U.S. This discourse of 'national hospitality' further works to conceal just how significant U.S. immigration policy has been in both the continuous creation of highly exploited (im)migrant labour, the restructuring of the American labour force and the institutionalisation of discrimination against 'foreigners' in the U.S. labour market. As Ali Behdad (2005: 26) puts it, '[s]uch a misrepresentation reaffirms America's ideology of exceptionalism, the idea that the United States is a free country, free of political oppression and religious persecution'.[7]

Such states of exceptionalism also ignore the racialized characteristic of the workforce made into 'guest workers'. Interestingly, in most of the debates, discussions of 'race' are absent. This is even more pertinent since the lack of such explicit discourse stands in stark contrast to the highly racialized discourse surrounding the entrance of those who are permanent residents or 'illegals'. Yet, it is not a coincidence that the vast majority of guest workers are nonwhites from the Global South, mostly from the Caribbean, Latin America and Asia (Meyer, 2006). Indeed, the long history of the racialization of unfree labour within the transnational space of capital and states is significant in understanding the indentured conditions imposed upon H-visa workers. The history of U.S. laws on immigration and citizenship illustrates how the state has always been actively involved in determining the terms of freedom for negatively racialized groups whose unfreedom has been a key source of profits for colonizing groups and subsequently for 'American employers'.

For migrants, the Naturalization Act of 1870, the Page Law of 1875, the Chinese Exclusion Act of 1882, the Scott's Act of 1888, the Gentlemen's Agreement between 1907 and 1908, the Immigration Act of 1917, the Quota Act of 1921 and the Alien Land Laws passed in 1913, 1920 and 1923, along with other legislative measures such as the 1850 Masters and Servants Act, are some significant examples of legislation that denied permanent residency and citizenship to nonwhites, restricted their employment rights and economic opportunities and reconfigured the occupational profiles of diverse groups of negatively racialized migrants from Asia, Latin America, the Pacific, the Caribbean and Africa (Chan, 1991; Hing, 1993; Boris, 1995; Andreas, 1994; Bannerji, 2003).

As Ian Haney-Lopez (1996) has shown, for negatively racialized migrants, state legislation on citizenship and immigration produced the formation of 'split labour markets' (Bonacich, 1972) by positioning nonwhites as noncitizens, propertyless and dependent on wage labour. These and other laws on citizenship and immigration had the expressed purpose of denying diverse groups of nonwhites the benefits granted to those racially organized as whites. For nonwhite migrants, this meant a denial of *permanent residency*. Significantly, their nonwhiteness was very much related to their status as unfree workers (Steinfeld, 1991; Linebaugh and Rediker, 2000). Moreover, their classification as temporary, contract labour made them ideal for short-term and flexible employment in *past* periods of capitalist expansion, indeed in the formative years of the transnationalization of capital (Glenn, 1986; Potts, 1990; Hing, 1993).

Today, the relationship between freedom, 'race' and ideas of societal membership remains central to the story of recruiting people as guest workers to the U.S. That such vulnerability is being organized precisely at the period in history when nonwhites have formal access to permanent residency in the U.S. in not a coincidence. Recall that after 1965, nonwhites admitted as immigrants, that is, as permanent residents, came to have (virtually) the same formal rights as white Americans and, of course, many *exercised* these rights. Furthermore, after 1965, a growing proportion of permanent residents came from the Global South. In the period leading up to the 1986 IRCA (that simultaneously made it more difficult to receive a green card and introduced the current H-visa program), a common conceptual practice of power in the U.S. was to organize the discursive problem of there being 'too many' nonwhites in the country. These too many nonwhites, it was repeatedly stated, were causing irreparable damage to the 'character of American society'. This continues today and is perhaps best captured in Peter Brimelow's 1995 book *Alien Nation*, which called for a reversion to pre-1965 U.S. immigration policies when legal preference was given to white, English-speaking migrants.

The H-visa program was one legislative 'solution' to this problem. This is because the racialized trope of there being 'too many' nonwhites was, in fact, a discourse that produced the idea that nonwhites living as permanent residents and citizens had *too many rights*. In the mid-1980s (not that distant from the civil-rights changes of the 1960s), simply eliminating these rights was not a viable solution. The H-visa system offered a way to legalize the *resubordination* of many nonwhites entering the country by recategorizing them as temporary and foreign workers and therefore denying them the rights tied to permanent residency and citizenship.

Of course these legislative changes also led to ever-larger categories of illegalised workers as well, since the numbers of green cards issued to 'employment-related' applicants became an ever-smaller proportion of the green-card system of immigration, and even these too were limited to a professional class of immigrants. This led to a return of practices that subordinated mostly nonwhites within the U.S. through the category of 'illegal' alien. With the 1965 changes, and even more so with the 1986 changes, pressures leading to the liberalization of racism, then, were met by a countertrend towards the greater restriction of the rights and entitlements of nonwhites in the U.S.

In this racialized economy of bodies, border-control measures enabled national states to reorganize their nationalized labour markets to *differently include* a group of guest workers rendered unfree, that is, legally tied to specific employers, and therefore enormously vulnerable to these employers' demands as well as to create a much larger group of illegalised migrants who, like 'guest workers', were primarily from the Global South. In the current historical juncture where both people's displacement and subsequent migration is occurring at historically unprecedented levels, a racialized sense of 'belonging' in the U.S., with its legitimisation of differential legal regimes for citizens and foreigners, remains a central motor force for capitalist globalisation. Encoded within contemporary U.S. immigration policy concerning workers, then, is the dominant logic of exploitation and power imbalance inscribed within the racialization of membership in the U.S.

CONCLUSION

Given the H-visa system along with the movement of millions of illegalised migrants, it is clear that restrictive immigration polices have not worked to restrict people's mobility. They have, however, restricted the ability of most migrants to get 'immigrant', or permanent, status. The actual material reality of imposing restrictions on immigration, then, is the creation and subordination of official nonmembers of Americanised society. Border-control efforts, therefore, are legitimated not through their effectiveness but through reference to another set of borders—racialized borders between Us and Them. These borders are materialized through national state practices and the 'sovereign right' of states to limit their membership. Such borders are legitimated through ideologies of nationalism that naturalize discrimination against those who can be classified as 'foreign'.

Therefore, the virulently racist, anti-immigration discourse currently raging in the United States and the increase in the numbers of people migrating to the U.S. as either 'illegals' or as guest workers should not be thought of as paradoxical or contradictory. In fact, the two need to be understood as complementary processes. Anti-immigration discourses, by continually constructing certain racialized and gendered bodies as always 'foreign', work precisely by maintaining their economic viability to employers (Hage, 2000). In fact, the denial of permanent status to the growing numbers of 'illegal' or guest workers, most from the Global South, has made them into the cheapest and weakest labour force in the U.S. As Hage puts it, 'they are best wanted as "unwanted"' (Hage, 2000: 135). In the process, two major goals of U.S. state rule are achieved: a subordinated and, therefore, cheapened work force is created, and an American national identity is reinforced through the othering (and criminalisation) of (im)migrants.

In this regard, the well-publicized 'debates' on immigration policy in the U.S. that the state and media help to organize through continuously commissioning and publishing polling data on what 'Americans' think of 'new immigrants'—that is, how many there should be, where they should come from, what status they should be given and so on—should not be viewed as 'a meaningful tool for the formulation of policy, but rather such immigration polls and debates should be see, in a more anthropological spirit, as 'rituals of *white empowerment*—seasonal festivals where white ['Americans'] renew the belief in their possession of the power to talk and make decisions about Third World–looking people' (Hage, 2000: 241).

In this sense, then, within this period of neoliberalism, the racialized 'nation' continues to serve as 'the ideological alibi of the territorial state'

(Boyce Davies, 1994). The U.S. labour market is able to be restructured along neoliberal lines, in part, because of the legality of the discrimination against both illegalised and 'guest' workers. Conceptual practices of citizenship in the U.S., therefore, can be said to act to continuously attach the citizenry to the governance of American society or, to paraphrase Hannah Arendt, to create that most potent of forms of self-governance—the national state or the state that is seen as wholly legitimate in ruling because many people imagine it as simply an extension of the popular will of The People to which they wish to belong.

A common theme in all policy changes I have examined has been the emphasis on shifting the multiple meanings of national membership in ways that help to organize and legitimise a global apartheid, an apartheid that, as in the past, is formed through simultaneous processes of racialization and nationalism. With the categorization of people as 'illegal' or as 'guest workers', the state quietly borrows from the exclusionary practices organized through concepts of citizenship and its ideas of the fictive national society, in order to reposition these subordinated (im)migrant workers as part of a 'foreign' workforce in the U.S. That some people can live, work, pay taxes and even die in a particular space but still be considered foreign to it is clearly an ideological representation of space and people's place in it.

All of this has been exacerbated and expanded since the events of 9/11. The disciplining and further criminalisation of the vast majority of (im)migrants to the U.S. is one of the cornerstones of the USA PATRIOT Act of 2001.[8] This act followed the Anti-Terrorism Act of 2001 to further empower federal security agencies to weaken the civil liberties of citizens while ensuring that the figure of the 'terrorist' was sutured onto that of the 'foreigner' or 'immigrant'. It also created new racialized and nationalized distinctions between citizens and noncitizens as well as between differently racialized citizens. Among its most signal achievements is the state's ability to indefinitely incarcerate noncitizens (Section 412) and deny all noncitizens the rights of habeas corpus and due process. Indeed, much of the 'antiterrorism' work done through the USA PATRIOT Act has been targeted at noncitizens. This has had a further chilling effect on the grassroots organizing ability of (im)migrants, some of whom, like South Asians in New York state, have experienced an enormous loss of community members, through acts of detention and deportation for minor immigration infractions.

Of course, (im)migrants) and their allies have resisted. Testimony to this resistance is the enormous, indeed historically large, marches and other demonstrations organized and held in 2006 primarily by (im)migrants (many of them illegalised) that challenged their legal, political and social subordination in the U.S. Across the country, tens to hundreds of thousands marched in New York, Washington, DC, Las Vegas, Seattle, Miami, Chicago, Los Angeles, San Francisco, Atlanta, Denver, Phoenix, New Orleans, Milwaukee and other cities, further showing the widespread presence of (im)migrants in U.S. society. Seen by many as the most powerful of contemporary civil-rights movements, many at these rallies were spurred onto action by the late 2005 passing of HR 4437 (the U.S. House of Representatives Border Protection, Antiterrorism, and Illegal Immigration Control Act that promised to turn illegalised (im)migrants into felony criminals), which they forcefully rejected. However, there were other, more radical demands made at these rallies as well. Chanting slogans like 'sí, se puede!' (yes, it can be done!), 'No human being is illegal,' and 'God knows no borders,' participants also sought a wholly new world not premised on some of the key aspects of modernity—national state sovereignty and the 'right' of states to limit membership and, therefore, target those seen as 'foreigners'. Demands included the equalization of treatment of all people in the U.S. and, therefore, an end to the selective privileging of those both legally and socially seen as 'American citizens'.

What was also evident at these rallies was the transnational character of the struggle. Reflecting the makeup of the population of illegalised persons in the U.S., many of the participants were Latino; however, people from across the world represented themselves and spoke of a common, global struggle against exploitation and displacement. Such movements demonstrate the multipronged struggle for both equality and for justice and are a clear response to the system of apartheid currently embedded within U.S. legislation and culture.

NOTES

1. I use the somewhat awkward term '(im)migrants' to remind the reader that the term 'immigrant' has specific legal meanings for those who carry this label. Immigrants are *only* those who are granted permanent residency rights in the U.S. (i.e. 'green-card' holders) with which comes a number of rights, protections and entitlements that all other people migrating to the U.S. are denied.
2. Kochhar (2007b: 16) notes that '[i]n 1995, 10 per cent of newly arrived Hispanic workers were employed in agriculture and 11 per cent were in construction. By 2005, only 5 per cent entered into agriculture and 25 per cent were hired into construction. Earnings in agriculture are known to be below average, and earnings in construction are above average'.
3. Having said that, however, it is important to note that the social processes of racialization fundamentally shape one's experiences of citizenship or its denial. Just as there are variances in the experiences of (im)migrants, the ongoing saliency of ideas of 'race' and practices of racism in the U.S. continue to ensure that nonwhite citizens—blacks, Latinos, Middle Easterners, Asians—are cheapened in relation to their white counterparts.
4. Discrimination is said to take place when 'individual workers who have identical productive characteristics are treated differently because of the demographic groups to which they belong' (Ehrenberg and Smith, 1994: 402).
5. The H-1 visa was initially created by the 1952 Immigration and Nationality Act (INA) for the recruitment of individuals 'of distinguished merit and ability' for temporary work in the U.S. The requirement that the job be temporary was eliminated in 1970 (Meyers, 2006).
6. Unfortunately, there are, as yet, no studies comparing the wage rates of comparable H-visa and illegalised workers.

7. The origin discourse of the U.S. being founded on democratic principles by a people fleeing persecution in search of freedom also works to disavow both the genocidal practices against diverse indigenous societies and the thefts of their lands as well as the enormous disparities in power and wealth between the early settlers.

8. Capitalizing this act is not a grammatical error but signals that the name of the act is an acronym: Uniting and Strengthening America by Providing Appropriate Tools Required to Intercept and Obstruct Terrorism Act.

8 Trade Unions, Migration and Racism in France

Steve Jefferys

France lies at the intersections of the Mediterranean and Atlantic and of the south and the north of Europe. In the last millennium it was consistently Europe's wealthiest country, and for 250 years it has been at the political epicentre of modernity: the historic home of human rights and individual freedoms. France's central location on the world's major trade routes did not just mean that most of the world's commodities crossed its territory: so too did a huge variety of the world's human beings. Migrants were pushed by economic or political repression and pulled by labour shortages, French economic prosperity and the prospects of political and cultural *liberté*. These migrants, among the most dynamic of their own generations, added and add enormously to France's already huge existing diversity of economic, cultural, artistic, intellectual, culinary and social life. Over the last fifty years, as the numbers of migrants from non-European countries to France have risen, this diversity has become much richer. But richness and diversity have not always been universally welcome. Trade unions, in particular, have always had a degree of ambivalence about migration. 'Foreign' workers have often been viewed as 'competitors' who would undercut 'local' wages and conditions.

This chapter sets out to examine how French trade unions responded and respond to this recent visible migration and to the more prevalent antimigrant political discourses of the last five years.[1] French trade unions have very low trade-union membership, with density currently estimated at around 8 per cent. But, thanks to legal rights to enterprise-level staff representation, with privileged access for the five politically distinct national trade-union confederations, and to their joint management with the employers of much of the French welfare system, French unions still have more legitimacy and mobilising capacity than in several other EU member states with much higher nominal trade-union membership (Jefferys, 2003). This mobilising capacity, and the presence of a 50 per cent level of confidence in the trade unions since 2002, was testified to again in the huge mobilisations in 2003 (unsuccessful) and in 2006 (successful) (Andolfatto, 2007).

The chapter first discusses the context of immigration into France. Next, it presents some national-level evidence of persistent discrimination in the French labour market and highlights some recent research findings from the transport, hospital and retail sectors. This evidence confirms the experience of racial discrimination at work, lived by many migrants or children of migrants, and suggests that some French trade unionists at workplace level are not responding well to this new challenge. It argues that the French republican approach makes it especially difficult for the French trade unions to combat such discrimination.

MIGRANT FRANCE

Millions followed in the migrating footsteps of the Swiss Jean-Jacques Rousseau and the German Karl Marx to settle permanently or temporarily in Europe's own 'melting pot' (Noiriel, 1986). In the nineteenth century some 300,000 Belgian workers provided the bulk of the manpower (with Germans and Italians) needed for the expanding mines and factory towns of northern France. It was the birth of modern nationalisms in the last imperialising quarter of the nineteenth century that turned 'local' rural suspicions of all 'foreigners'—from other villages as well as countries—into 'national' distinctions and forged an ideology of 'racial superiority' (Tripier, 1990). Xenophobic reactions to the increased use of 'foreign' labour became common: between 1867 and 1893 there were sixty-seven recorded accounts of protests directed against Italians, eleven against Belgians, seven against Germans and two against Spanish workers. While the eighteenth-century workers' passbook dropped out of use for 'national' workers at that time, France still required them for 'foreign' workers right up to the First World War—an early echo of the current system of work permits.

Migration continued after the Great War, but whereas before they came predominantly from geographically close 'poorer' neighbours, in the 1920s 1.5 million migrants came from Poland and Southern Italy. By 1931 foreign workers made up 42 per cent of French miners, 30 per cent of building workers and 38 per cent of engineering workers, and it was they who bore the brunt of anti-immigrant reaction in the 1930s with the 1934 conservative government promising 'priority to French workers in the job market' (quoted in Cross, 1983).

After World War II, between 1955 and 1975 1.7 million migrants entered France: Portuguese and Algerians entered France in roughly equal proportions, followed by Spaniards, Italians and Moroccans (Marchand and Thélot, 1991). Sometimes their rates of pay were below those of 'French' workers when doing the same work because they could be put on a lower classification through lack of 'training'. But almost always the jobs could be paid less and, because they were dirtier and harder, were the jobs that the 'French' preferred not to take. France closed its frontiers to new permanent immigrants in 1974 and initiated an aided-repatriation programme in 1977. Subsequently, under right-wing governments in 1986 and in 1993, it first made it easier to expel 'undesirable immigrants', and then denied automatic citizenship rights to children born in France or to any 16- to 21-year-olds who had received prison sentences of six months or more, even if they had lived their whole lives in France (Schor, 1996: 280–2). As migrants sought other means of entry, the rate of political asylum rejections skyrocketed: up from 15 per cent in 1980 to 84 per cent in 1990 (Schor, 1996: 277). In the single year 2000, the numbers held for an average of 5.1 days each, in thirteen of France's twenty-three immigrant detention centres, was 17,883, a rise of 25.4 per cent over 1999, with nearly half (44 per cent) subsequently being deported (*Le Monde*, July 27, 2001).

Nonetheless, migrant numbers still continued to grow through family reunion, education and asylum provision: over the ten years 1994 to 2003 there were 382,000 requests for asylum in France while in 2002 the numbers of 'immigrants' registered for legal residence (including students) rose to 206,000, the highest total since 1974 (Thierry, 2004). By 2004–5 metropolitan France hosted 4.9 million people born outside the country, some 40 per cent of whom were naturalised, and an estimated additional ten million children and grandchildren of immigrants. In 1990 the proportion of the French population born outside France was 7.4 per cent and it rose to 8.1 per cent in 2004–5 (Borrel, 2006). This was not very much higher than the previous peak of 7.0 per cent in 1931 (Marchand and Thélot, 1991), but the contemporary experience of migration is very different from the pre–Second World War one.

Since 1950 migration into France has been more visibly distinct from the native population; it is often more concentrated geographically in particular 'suburbs' or the *banlieu* that in the twenty-first century became a racialised geographical synonym for localities with high proportions of people whose origins were Arab or African; and it is more equally balanced between men and women.

The French labour market has also undergone rapid change since the boom years of the 1960s and early 1970s: the number of manual jobs has declined rapidly, and unemployment has risen. Between 1975 and 1990 as many as 40 per cent of the jobs occupied by non-French citizens disappeared: the equivalent of half a million redundancies. Within increasingly depressed *quartiers* or high-rise estate *cités*, the descendants of North Africans increasingly became a dispossessed or excluded population working, where they found work, in more marginal and lower-skilled jobs. By 1996, 24 per cent of all temporary workers in France were foreigners, and three-quarters of these came from Algeria, Morocco or Tunisia.

These migrant workers were increasingly victimised by an increasingly racist political discourse aimed at winning the votes of other low-skilled French workers who felt at risk in an increasingly insecure labour market. Thalhammer et al.'s (2001) Eurobarometer, study based on roughly 1,000 interviews in each country, categorised people in terms of their responses to six questions as being actively tolerant, passively tolerant, ambivalent or intolerant. This 'intolerant' category captured people who:

feel disturbed by people from different minority groups and see minorities as having no positive effects on the enrichment of society. They have a strong wish for assimilation. Furthermore, the intolerant support the repatriation of immigrants and the very restrictive acceptance of immigrants. Intolerant people tend to be less educated and less optimistic (according to their personal situation) than the average. (2001: 24)

Table 8.1 shows the results carried out in France and three other countries, and compares them with the EU15 average. At the time of the survey, in 2000, while France was not the most intolerant EU country, 45 per cent could be described as 'ambivalent' or 'intolerant' towards immigrants compared to 39 per cent across the EU15 as a whole.

The presence of a significant minority of French citizens ready to indicate their lack of toleration, understanding and sympathy for immigrants was reflected in the first rounds of the 1995 and 2002 presidential elections, when 30 per cent of all manual workers voted for the racist *Front National* candidate Le Pen (Courtois and Jaffré, 2001). In the seven-year interval between the two presidential elections the proportion of the unemployed who voted for him rose from 25 to 38 per cent (*Le Monde*, April 28, 2002). It is also clear that the level of both general xenophobia and specific anti-Arab and anti-Muslim sentiment rose in France following 9/11 in New York and its aftermath. Migrants and the descendants of migrants experienced this acutely. An Algerian-origin bus driver explained to French researchers for the RITU EU-funded project[2] in January 2004:

Everything that happens on the TV. I'm going to tell you the truth. It's making people crazy: Israel, Iraq, all that . . . People have had enough. Secularism, everyone's talking about it everywhere. The drivers, they're talking about it. The whole world's talking about it. You can't get away

Table 8.1 Attitudes Towards Minority Groups (Ranked by Per Cent Intolerant), 2000

	Intolerant	Ambivalent	Passively Tolerant	Actively Tolerant
Belgium	25	28	26	22
France	19	26	31	25
UK	15	27	36	22
EU15	14	25	39	21
Italy	11	21	54	15

Note: Differences of 6% and more are statistically significant.
Source: Thalhammer, Zucha et al., 2001).

from it. We're right in the middle of it. We don't give a damn but, whatever, we're in the debate, so . . . I feel that's what affects people, what's on the TV . . . These things happen thousands of kilometres from here but apparently if I stole something my father would cut off my hand, perhaps, what.

A November 2005 opinion survey, undertaken in the wake of the then Interior Minister Nicholas Sarkozy's description of young minority people in the suburbs as 'scum' and the nationwide car burnings that this provoked, found one-third of the French interviewed describing themselves as 'racist'— up from one quarter in 2004.[3]

With the subsequent election of Sarkozy as president in 2007, French xenophobia thus moved from its confinement within the extreme right to become an official part of the centre-right political agenda. Despite appointing two women of North African origins as ministers of justice and urban issues, in his appointing a 'Minister for Immigration, Integration, National Identity and Joint Development', and moving the political asylum function to the new ministry from the Foreign Office, the president confirmed his view of immigration as the source of a challenge to national identity. Brice Hortefeux's policy is to be hard on 'illegal' migrants, to restrict immigration to certain chosen occupations where there are labour shortages, but to 'protect those (immigrants) who share our rules and our values' and who are ready to assimilate into French society (*Le Monde*, 9.11.07). The minister of immigration now aims to increase deportations to 25,000 a year (*Le Monde*, 1 June 2007) and boasted of achieving 20,000 exits in the first six months of the year (*Le Monde*, 9 November 2007). Already a new nationality law in 24 July 2006 had repealed the entitlement to automatically claim French citizenship after ten years' residence. The fifth immigration law in five years, passed on 24 October 2007, added several 'precisions' attacking family reunion: any 16- to 65-year-old person seeking to enter France as a spouse, parent or child must now pass a national identity test in French values and the French language, and the host family must prove it has the resources to support the incoming family member. It also sanctioned a DNA-testing procedure in case of doubts. In addition, the 2007 law makes the asylum process more difficult, by requiring anyone whose request for asylum is rejected to appeal within forty-eight hours. This legal tightening up on what in reality is 'non-European' immigration is in a context where there is already considerable labour market discrimination against existing ethnic minorities but also very little done to combat it.

LABOUR MARKET DISCRIMINATION AND THE LAW

In French law it is unconstitutional to ask people to self-classify themselves into different ethnic groups, but it is possible for surveys and the census to ask people to identify their nationality and their country of birth and that of their parents. Using this information, one study showed that in 1998, while male unemployment among native-born French aged between 18 and 40 was 10.1 per cent, it was 24.9 per cent among Turkish immigrants and 29.6 per cent among Algerian immigrants, and was still 21.2 per cent and 23.2 per cent among second-generation immigrants. In a second publication based on the same 380,000 survey responses, the authors go on to show that while second-generation working-age descendants of Portuguese, Italian and Spanish immigrants have the same unemployment risks as the descendants of French-born citizens, both the risks of unemployment and of being employed in insecure work remain much higher for the descendants of non-European immigrants from North Africa, sub-Saharan Africa or Turkey. The data confirm that this discrimination also operates in the public sector: not only are there disproportionately fewer second-generation visible minorities employed in the public sector, but where they have found work, it is in lower quality public work (Meurs, Pailhé et al., 2005).

For several years a number of organisations have undertaken discrimination testing in France, comparing the chances of French nationals with 'majority'-sounding names and 'white' skins, with those with Arab or Black African-sounding names and skin colours. The most recent and rigorous series were conducted in an ILO study involving 1,100 useable tests from six urban areas (Lille, Lyon, Marseille, Nantes, Paris and Strasbourg) and focused on retailing and sales, construction and hotel and restaurant and professional employment sectors (Cediey and Foroni, 2007). This found that at the recruitment stage the behaviours of just over half of the employers involved were clearly discriminatory: 33 per cent practised discrimination based on the name and presumed origins of equally qualified and otherwise identical candidates; 11 per cent discriminated in favour of the 'white' candidate when deciding to put him or her on a waiting list; and a further 7 per cent opted for the 'majority' ethnic candidate at the interview stage. Overall, when an employer had to choose between a 'majority' male or female candidate and the directly matched equivalent French national whose roots were in North or sub-Saharan Africa, nearly four out of five times they chose the 'majority' candidate. The 'best' or least discriminatory result was for Arab women: employers would only discriminate against them two-thirds of the time. The evidence of discrimination at the point of entry into the job market is overwhelming, as shown in Table 8.2.

Similar results to this were reported to our RITU researchers in a public sector hospital. An African-origin male hospital worker explained what happened when a woman married to a North African applied for a medical secretary post:

> One of my mates whose surname is 'Ben X' sent her application here. Well, no, there was no job. It's a girl, see. She's married, and her maiden name is Martin. She then sent in the application under her maiden name

Table 8.2 French Discrimination Testing Result, 2006

When French Employers Favour the Ethnic Majority Male or Female Candidate		
Four Out of Five Times	*Three Out of Four Times*	*Two Out of Three Times*
over a black African woman		
in hotels and restaurants		
	over a North African man	
	in retailing and sales	
		over a North African woman

Source: Cediey and Foroni (2007: 110).

and then she was taken up right away . . . After, as soon as they have her on the telephone, they will hear her accent . . . There are several stages in recruitment. Your name, something as simple as the fact of having an immigrant name. Today that can be a handicap to getting a job. I know it. I've already seen it lots of times.

Despite such overwhelming evidence, successive French governments moved only very slowly to transpose the European Equal Treatment (racial or ethnic origin) Directive (Directive 2000/43/EC), which was supposed to take effect from July 2003 and which marked an important advance in terms of understanding that racism is both direct and indirect. While its Article 11 explicitly refers to the Social Dialogue, enjoining Member States to:

'take adequate measures to promote the social dialogue between the two sides of industry with a view to fostering equal treatment, including through the monitoring of workplace practices, collective agreements, codes of conduct, research or exchange of experiences and good practices'.

The initial French transposition of 16 November 2001 just amended the existing labour code article combating discrimination. It was named very generally the law 'concerning the struggle against discrimination'.[4] Trade unions were given the right to take legal proceedings on discrimination grounds without having to prove they have a mandate from an employee, and in future all sector-level collective agreements would have to contain an antidiscrimination clause if they were to be extended by the Ministry of Labour. The government also committed itself to establishing a free victim telephone support line. However, although the law introduced the concept of 'indirect discrimination', not only was no definition given, but in the pre-report on the law the definition that was given had much greater ambiguity than in the directive. It described 'indirect discrimination' only as a long-term disadvantage affecting workers collectively: 'Apparently harmless actions whose repetition and accumulation lead over several years of practice to major differences between whole groups of employees'.[5]

At worst this definition deliberately deprived individual workers of any protection against indirect discrimination, and at best it reflects the absence of any serious reflection on what was an entirely new concept in France.

The original directive's concept of 'indirect discrimination' clearly challenged the traditional French 'blindness' to the societal structures of racist discrimination (Vourc'h and de Rudder, 2002). This ambiguity was accompanied by a very long delay in establishing any form of body for the specific promotion of equal treatment. It was only in February 2004 that the reelected French president, Jacques Chirac, announced the setting up of an Equality Authority (the High Authority Against Discrimination and for Equality, HALDE). This was required under the directive's Article 13, but the law establishing it was actually only passed on December 30 2004 with the decree and the nominations of its leading committee taking place in March 2005. Its responsibility was to 'struggle against all discriminations: skin colour, geographical or social origin, age, gender, sexual orientation, disability, political and religious opinions',[6] and its nominated president was Louis Schweitzer, the former chief executive of Renault. Two of its other ten members were Cathy Kopp, the head of human resources at Accor, and Nicole Notat, the former general secretary of the CFDT. Two trade unionists, Odile Bellouin (CFDT) and Michèle Monrique (FO), were also nominated to the initial 18-strong consultative committee that was finally set up in September 2005. Subsequently, after the nationwide urban riots of October–November 2005, an 'Equal Opportunities' law of 31 March 2006 was passed by the Villepin government. This gave individuals the legal right to carry out discrimination tests that could lead to a successful prosecution. But this law was largely rendered inoperative following mass demonstrations and strikes against the clauses introducing exemptions from labour protection of all workers aged under 26, and with their repeal, the proposal to give the antidiscrimination body some teeth disappeared. In 2006 HALDE received 4,058 complaints of discrimination, with the biggest (35 per cent) concerning discrimination related to the complainants' 'origins', yet it could report only two legal cases that year in which judgements were made against 'origin' discrimination in recruitment (HALDE, 2007).

Despite the paucity of legal action against discrimination related to workers' origins in relation to work, and despite the absence of statistical sources clearly capable of demonstrating the ethnic penalties experienced by first, second or third French workers originating in the Middle East, Far East, sub-Saharan Africa or North Africa, there is a growing body of qualitative evidence illustrating this reality.

RACISM AT WORK IN FRANCE

The research undertaken by the RITU project detected four different forms of direct racism in French workplaces: racism through recruitment, racism through explicit discriminatory rules, racism from work colleagues and racism from customers.[7] Yet indirect racism was even more widespread. This is where the outcomes of organisational procedures and individual behaviours are to systematically make access or promotion more difficult for those whose 'origins' or skin colour are not those of the majority of French nationals.

The first form of direct racism occurs when black candidates are rejected purely because of their country of origin. This works in two ways. Restrictions barring non-French and non-EU citizens having permanent posts in the public sector directly disadvantage tens of thousands of migrant workers who have spent their lives working as temporary workers in the French public sector, and whose careers were blocked and pension entitlements reduced. But it also operates against many French citizens from the Overseas Departments. This is because French citizens born overseas and employed in the public sector have rights to special paid return holidays that mean they cost more than their French-born children or other black workers not born in the Overseas Departments. Sometimes the question is directly posed to the workers in their recruitment interviews. A black woman health-care assistant told a RITU researcher:

> When I got to [the Paris hospital] Saint-Louis for a recruitment interview, they asked me: 'Were you born here or there?' 'I was born here'. 'Oh good'. Right, that was positive. You see, if I'd told them that I was born there, no thank you.

Direct discrimination in recruitment was being openly used here against West Indies–born black workers who were French citizens. Racist discrimination against all ethnic minorities in recruitment, however, is still common. One hospital clerical worker described how in her office the new boss had moved out the three black staff, and never converted the temporary North African–origin worker replacements into permanent posts: 'If it's ever "Magrehbins", it is only for short-term contracts. As far as permanent jobs are concerned, you'd think those people never applied'.

A second way in which direct racism surfaced in France is in the imposition of discriminatory rules at work. Head scarves are prohibited, for example, by Carrefour works rules at one large hypermarket in the South of France. The rules ban 'ostentatious' religious or political dress and are endorsed by the trade unions. When the store was opened there was also an initial ban on ethnic-minority staff talking to customers in their own language. The workers saw this both as a direct form of discrimination and as making no sense, since these language skills were a positive resource:

> To start off with . . . what they told us was that we weren't allowed to speak our own language. But I didn't take any notice, because when you got these old gents coming in who really couldn't speak French, but came to ask us something.

Talking in their mother tongue heightens management's suspicion of collusion between staff and customers, while workers perceive the ban as an additional sign of being discredited and even criminalised (as potential conspirators).

A third area of direct racism occurs in the constant bullying by some fellow workers or when nonwhite majority workers are exposed to particularly critical attention by supervisors and managers. In one French hospital a survey picked this up, with one in ten workers indicating it was for racist reasons.

A SUD Health Federation activist told another story of this background racism, where hospital staff make racist comments about a young North African patient who had fallen from his scooter:

> And well, in the preparation room behind his back, what was said, was: 'Didn't he fall because he wanted to get away from the police?' The second thing was: 'Hadn't he stolen the scooter? All that because he was a North African (maghrébin). If it had been a White there wouldn't have been these reactions. We are living in a very strong dose of latent racism that comes from a few, but which often poisons us all.

Finally, several of the French interviewees highlight direct racism from the customers. This was the most common form of racism reported. Examples were given of patients who said '*I don't want a Black to touch me*'. One interviewee in retailing indicated that nonwhite checkout staff were more likely to be on the receiving end of aggression. A French 22-year-old North African bus driver in France gave the most common illustration:

> At the beginning, when you are a driver, during the first months . . . the little granny you pick up at C. over there (name of a middle-class area of the town). When she gets on and when she sees the driver . . . it isn't even worth it for her to talk, it's on her face, eh, it's written, it's printed . . .

The visibility of nonwhite workers dealing with the 'public' creates a background racial tension for many of them as they detect nonverbal hostility even in the customer's body language. Survival means simply getting used to it.

Indirect discriminatory treatment as a result of a person's origins is even more widespread in France than direct racism, even if it is very often denied. This is clear within the French hospital system, where there is a strong ethnic hierarchy operating, with French whites at the top. There is a marked concentration, through recruitment processes of first-, second- and third-generation French citizens with overseas origins, in manual and clerical jobs. But these then are precisely the jobs—as health-care assistants and administrative and maintenance personnel—that are the main targets of hospital cuts, contracting out or casualisation. Hence, the norms of the 'racially neutral' evolution of the health-care labour market actually lead to still greater disadvantage being heaped on ethnic-minority workers. Those who are already near to the bottom of the occupational pyramid are then the most exposed to restructuring and often arbitrary management. The simple fact that these are largely ethnic-minority workers effectively reinforces indirect racism. At the same time, the very strong tradition in France of moving between classifications through the mechanism of sitting internal exams, often in maths and French, tends to reinforce the stratification of the labour market. One senior hospital-level trade unionist illustrated how even getting workers up to the right level could pose problems:

> One worry in our hospital is that it's the head of personnel who helps get people up to scratch. Courses of French and maths, and most of the staff don't want to go because it's her . . . So the people don't do the revision, they go straight to the test and fail it because they're not ready. Lots of our West Indian colleagues are in that boat . . . There it's free, because she does it in her working time. But if you bring in a teacher, you've to pay them.

Still other Overseas Department–origin interviewees reported that their managers either didn't tell them about the exams or put the notices up at the very last moment, not giving them time to prepare properly.[8]

The suggestion that what is taking place is indirect, structural racism is confirmed at the Carrefour-Grand Littoral hypermarket at Marseille. There, although it has had a locally focused recruitment policy of ethnic-minority junior staff ever since it opened ten years earlier, there are still no minorities in senior company or trade-union positions. At first Carrefour tended to keep the minorities who were recruited out of direct contact with the public. Today, labour shortages and the changing composition of the client base have reversed that informal policy, and French nationals from North African origins are now the main visible hypermarket labour force, but the warehouses still have a high proportion of workers with sub-Saharan African origins. One male North African trade unionist explained the continuing lack of 'people from foreign origins' in top jobs as a problem of 'French society':

> I don't know whether that comes from the company or the union or what, whatever it is that prevents people from foreign origins from getting on. My view is that it rather comes from French society that isn't perhaps yet . . . Me, I make the comparison with other countries, Great

Britain or the USA, where you can meet high-placed people coming from foreigner backgrounds. There isn't any . . . a priori about it, what . . . It's true that in France there isn't . . . It's French society that isn't ready to accept this type of . . . development.

This experience is not just confined to the largest private sector employer in France, Carrefour. One 25-year-old North African bus driver told us why you don't have to wonder why more blacks have disciplinary problems than whites. The simple fact is, he told the white interviewer: 'There is you and there is me. You are going to fail, but it's ok, we are going to make it be fine. But if I fail, they're gonna make a fuss . . .' The same rules apply universally in France, only they tend to get applied more stringently by line managers to black workers than to workers from French-majority origins. This is clearly institutional racism.

FRENCH TRADE UNIONS TODAY

In France all five major union confederations now have long-standing anti-racist policies, standing for equal rights for all, despite having been previously associated with calls for strong state controls on immigration in times of recession. This dualism gives rise to considerable ambiguities, even if the unions are now broadly favourable to the 'integration' or assimilation of minority workers into the movement (Bataille, 1997). The CGT 1995 constitution's opposition to racism is typical of the others:

Through its analysis, proposals and action, the CGT . . . campaigns for a democratic society that is free from capitalist exploitation and other forms of exploitation and domination, against discrimination of all sorts, racism, xenophobia and all forms of exclusion.

But these official antiracist policies at the top have major problems in articulating downwards to the base. This was clearly reflected in the significant proportions of sympathisers of the main confederations who voted for the two extreme-right candidates in the first round of the presidential elections of 2002.[11] Of course the CGT, but also the CFDT, FO and the much smaller Christian confederation, the CFTC, all associated themselves strongly with the transformation of the 2002 May Day demonstrations into a one million–strong protest against Le Pen and racism, just days before the second round of the presidential election. And both of the two largest confederations have formal positions opposing the increasing numbers of checks on identification papers, and both call for equal access to social-welfare benefits for both European and 'third-country' nationals (Lloyd, 2000: 118).

The two main French unions also now report a range of antiracism activities: campaigning for the removal of nationality clauses impeding the hiring of foreign workers (CFDT and CGT); mobilising to prevent the *Front National* expressing itself within work (CFDT and CGT); supporting the struggle of undocumented workers; signing up to sector agreements proposing training for young people from deprived areas (CFDT); educational campaigning against the confusion between insecurity and immigration (CGT).

In November 2004 the confederations demonstrated together around the call 'Racism threatens workers' solidarity', and in January 2005, the CGT, CFDT, CFTC and UNSA signed a trade-union charter *For Equal Treatment, Non-discrimination and Diversity*. This referred to the ETUC's antiracism programme and committed the signatory confederations to promote 'diversity' in the world of work as well as within their own organisations. The French charter, however, also specifically rejected the idea of a 'quota' as likely to create new forms of discrimination, arguing instead for the importance of strict equal treatment. A CFDT minority origin official described the policy as follows:

> It's not a matter of treating Mohammed more favourably, but of making sure that neither Mohammed, nor Rémi, nor Françoise, nor Sarah, nor Fatima, are excluded from the recruitment process because of their supposed or real origins.[11]

The evidence we provided earlier, however, suggests that this may not be sufficient to eradicate discrimination based on racial stereotyping.

The CGT, which has the highest level of support among French trade unions in works committee and industrial tribunal elections, insists that antiracism is a part of the general trade-union struggle. For example, in a leaflet aimed at recruiting minority workers it calls for nondiscrimination in recruitment and promotion, as a contribution to the creation of a proper lifelong employment status for all workers, and in March 2006 it produced a general poster 'Against racism and exclusion; for equality of rights'. Racism is analysed by the CGT primarily as a weapon used by employers to divide workers, so their argument is that only mobilisation can shift the balance of forces between employers and unions to counter it effectively. Coming from the French republican tradition, the CGT tends to draw a distinction between racism, which it sees symbolically or ideologically as any distinction drawn on the basis of appearance, and discrimination, which it analyses as disadvantage created by such distinction. This position tends to suggest racism is entirely the employers' responsibility, but the confederation has been aware for many years that some of its problems with recruiting and maintaining membership among 'foreign-origin' workers stemmed from the actions of the union's own members and activists, and, around 2000, it commissioned some research on the integration of minorities in two of its federations. However, more recently the CGT's 2004–6 Executive Committee still only had three ethnic-minority members out of 50, and before the 2006

Congress it expressed the hope it might improve its connection with French migrant workers by improving this proportion, while still not introducing any forms of quotas or reserved seats.[12]

The CFDT, France's largest confederation by claimed membership, argues that its activities in training, meetings and surveys are making its activists more sensitive about the need to elect minority representatives. Over the past five years it gave antiracist training to 500 of its leading trade unionists and, within an ESF-funded Equal project between 2002 and 2005, it worked in six regions to improve its antidiscrimination practice. Its position is that there is nothing inherent in capitalism that leads it to practise racial discrimination, so its strategy is to work with the employers to reach agreement on nondiscrimination measures.

More conservatively still, the FO confederation replied to a review of its antidiscrimination practices that since it is 'naturally' representative of French society it doesn't have to take any action to improve the representation of minorities. This was virtually the same answer as given by the CFTC, which also considers it already has 'considerable ethnic diversity within its internal structures'.[13]

DENIAL OR INTEGRATION

If the positions of the French confederations were firmly, if rhetorically, antiracist, there was a wider variety of responses by trade unionists at workplace level to the ongoing reality of racism. A CGT transport union leader acknowledged that local activists might see things differently than the national leadership:[14]

> During debates within the [national] union, we don't have any problem with principles. But if the others, in depots, considered they're not important, they are not going to apply them.

Essentially the RITU researchers found that in the health, retailing and transport workplaces studied, the trade unions had two main responses to the evidence of racism in the workplace: they either denied it, or, where they did recognise it and sought to respond, they tended to do so largely by making efforts to encourage 'foreign-origin' workers to integrate or assimilate into their existing structures, practice and policies. They were much less concerned to focus on the specific complaints of racism raised by the minority workers.

Denial occurred in several ways. In some cases there was a total failure on the part of the trade unionists close to the workplace to accept that 'foreign-origin' workers experienced racism. Asked about discrimination within the union, a white French CFDT health union official explained there had never been any discussion over how to integrate minority workers in the union because it was not a problem:

> To my knowledge at federal level, the question did come up once. Over the 14 years that I've been here, I've been alerted once to a problem, and that was settled by a telephone call. Political questions may come up, but the problem of someone's origin never does.

An ethnic-majority CGT woman health-union official, in describing how 'that kind of thing' was less likely, she thought, because it was a largely feminised workforce, made it clear that her definition of racism as was 'direct', in-your-face hostility. She'd seen the racist looks her black son-in-law got on one occasion in the street: 'Those looks, those . . . that yes! But . . . inside a hospital, including those with militants who could be black, walking around at Henri Mondor or in other places, that never'.[15]

Similarly, a French ethnic-majority retail worker reported: 'I've worked for two years at the bay, I've been around the drivers every day, and frankly I haven't heard any of that, you don't hear that kind of thing'. If racism is defined as being 'that kind of thing'—meaning open racist abuse or hostility, or a refusal to work with someone—then all kinds of racist discrimination that is not in-your-face can easily pass unnoticed. And if a client racially abuses or attacks a hospital worker or a driver, then, as a white bus driver who was a CGT regional union representative explained: 'A black man had been attacked, but for me it wasn't about race . . . I think it could have happened to anyone'.

The reality that this analysis ignores is that such 'incidents' do systematically occur more regularly to minority workers because of their 'foreign origins'. And such denials also occur where trade unionists know there are real problems. Thus a French CFDT retail sector official at Carrefour tacitly recognised there probably are some 'racism-related' problems: 'No racism-related problems get back to me. Either activists avoid tackling them, or things are covered up, and out of sight'. In the relatively exceptional circumstances of the French Overseas Department workers with a right to a paid trip home every two years it is more difficult to cover up such discrimination. Yet one of the white SUD hospital activists we interviewed sought to suggest that it was not racist discrimination involved when managers wouldn't take on these workers:

> I know certain trade unions, particularly the CGT, would say that some of the managers are racist because they don't want to take . . . because they pull back when they know that one of them (a French West Indian) is going to start with them. But I'm going to give the manager's point of view: if I take this person, me, every two years I've to pay for a special holiday.

This activist was creating a category of 'justifiable', nonracist discrimination but was immediately criticised by a woman SUD activist from the same hospital, who argued that the same logic could be used to discriminate against pregnant women.

The reasons for these direct and covert denial responses were usually either that the local union representatives were afraid of openly raising the issue because they felt that to raise the 'race' issue would harm interpersonal relations in the workplace, or that to do so could lose the particular union or union activist 'majority' worker support. In France the proportions of votes received in workplace elections directly determine how many representatives each trade union is allowed to have, and how many seconded-delegation hours they get. Thus, the presence of voters, particularly if the ethnic 'majority' workers make up the local majority, can create tactical denial.

Almost all the local union activists we interviewed in France who did not deny the presence of racism also believed that it would be wrong for the union to take any special measures either to encourage minority recruitment to jobs, or to encourage minority involvement in the trade union.

But more commonly without the stereotyping language, this insistence on universalism—don't give anyone any favours—is particularly strong. The French republican ethos starts from the normative assumption of equality and so dictates a 'colour-blind' approach that imposes a total assimilation to the French 'majority' by the 'foreign-origin' minority groups. Small accommodations may be made to the Muslim workforce, but larger ones will not.[16] Thus, one French CFDT retail federation official argued:

> A little stupid thing . . . We ask the restaurants never to serve pork. We don't even ask whether the comrades are practising or not. We think it's a way of respecting people . . . it's a way of integrating.

But otherwise 'integration' is a one-way street: any attempt to raise specific demands related to indirect racism and overall discrimination are rejected as likely to fragment the otherwise 'united' trade-union movement or working class.

The fact that this assimilationist approach generally does not work is illustrated by the extremely low levels of participation by 'foreign origin' activists in leading positions in the French trade-union movement—as well as by the lack of struggle against indirect racism in French workplaces. In the Paris RATP, where the unions largely operate between denial and assimilation, a few antiracist activists (from both majority and minority backgrounds) got together in 2003 to start campaigning to address the systemic racism that exists. One way of doing this, they proposed, was through an antidiscrimination charter. A North African activist argued it was important to involve the drivers and passengers and not to have such a charter simply be dictated by the company. In another illustration of positive action, one

SUD hospital branch 'adopted' two undocumented workers (*sans papiers*). This action, at one level purely political (and perhaps rhetorical), is a clear attempt by those involved to express their solidarity with a weak and much-maligned ground of migrant workers, and to give a lead on societal goals to the wider workforce.

CONCLUSION

This chapter has shown that immigration into France has been massive, and that many migrant workers and their descendants who appear to have 'foreign origins' through the colour of their skins experience significant discrimination at work. It has also suggested that while French trade unions produce antiracist political propaganda and in some cases a little training at the national level, they are often in denial about racism at workplace level. Where the unions do not deny its presence, they still tend to argue against taking any special measures against it. The highly political tradition of French unions (and their low membership density) means that they constantly claim to be representative of *all* workers in society and so see responding to the claims of a minority of workers, and in particular, they see the self-organisation of that minority, as equivalent to a breakdown in the workers' unity that is essential for mass mobilisation.

However, this strategy of proclaiming equality without struggling for it in those locations where unequal treatment is the norm has largely failed. If the anger of the second and third generations of French migrants of 'foreign origins' at the continuing racial injustice they face is to be harnessed by the unions as a part of their struggle against all social injustice, then French unions will have to develop a real inclusiveness to their way of operating and thinking from the bottom up.

NOTES

1. The chapter focuses on the experiences of visible minorities of migrants and their descendants who experience racial stereotyping through their external appearance or what is euphemistically called 'non-French origins' rather than the shared but also qualitatively different experiences of xenophobia towards Eastern European migrants to France since 2004.
2. The research basis for this chapter is largely drawn from the EU DG Research Framework Five 2003–2005 project, RITU, coordinated by the author from the Working Lives Research Institute at London Metropolitan University. The acronym means: Racial and ethnic minorities, immigration and the role of trade unions in combating discrimination and xenophobia, in encouraging participation and in securing social inclusion and citizenship.
3. A CSA survey of representative sample of 1,011 people interviewed face to face between 17 and 22 November 2005. *Le Monde*, 22 Mars 2006: 12.
4. LOI no 2001-1066 du 16 novembre 2001 relative à la lutte contre les discriminations.

5. «Mesures anodines dont la répétition et l'accumulation conduisent au terme de plusieurs années de pratiques à des différences profondes entre des groupes entiers de salariés» in VUILQUE P. Rapport fait au nom de la Commission des Affaires culturelles, familiales et sociales sur la proposition de loi relative à la lutte contre les discriminations. 10/10/2000. Cf http://www.assemblee-nationale.fr/.
6. Catherine Vautrin, Secretary of State for Integration and Equality, October 4 2004, http://www.premier-ministre.gouv.fr/information/actualites_20/.
7. Those conducting the interviews in France for the RITU project in 2004–5 included the current author, Christian Poiret, Véronique de Rudder, François Vourc'h, Philippe Poutignat, Christian Rinaudo and Jocelyne Streiff-Fénart.
8. C. Poiret (2005), French Health sector report, URMIS.
9. De Rudder et al. (2005), French retail Sector Report, Paris: URMIS.
10. Liasons Sociales, 29 April 2002, Briefing: CFTC: 25%; FO: 18%; CGC: 14%; CGT: 13%; CFDT: 12%; SUD: 3%. Liasons Sociales/CSA nationally representative survey of 5,352 voters.
11. Omar Benfaïd, CFDT Secretary, DARES Colloque, 9.12.2004.
12. Finally this aspiration did not apparently materialise. But since the CGT does not list its members by their ethnicity the author has not yet verified this.
13. The positions were reported by Novethic, 14.3.06
14. Poiret et al. (2005), French Transport sector report, URMIS.
15. Poiret et al. (2005), p. 20.
16. De Rudder et al. (2005).

9 Recent Migrants in the Canadian Labour Market
Exploring the Impacts of Gender and Racialisation

Valerie Preston and Silvia D'Addario

A country that recruits close to 250,000 newcomers each year, Canada currently favours skilled migrants in its selection policies. Approximately half of all migrants[1] admitted annually are skilled workers and their immediate families who are selected on the basis of each applicant's years of education, work experience and fluency in one of Canada's official languages (Citizenship and Immigration Canada Act 2006). In contrast, refugees and their families account for less than 15 per cent of all people admitted as permanent residents. The emphasis on skilled applicants was codified in recent legislation that explicitly separated policies regarding refugees from those directed at all other applicants for permanent residence. The Immigration and Refugee Protection Act, approved in 2002, specifies that refugee policies should recognize the humanitarian needs of those fleeing persecution and danger and ensure that Canada fulfils its international obligations, whereas immigration policies are 'to permit Canada to pursue the maximum social, cultural and economic benefits of immigration' (Canada, 2001, 3(1)a).

The weight placed on education and skills as criteria for admission to Canada has not translated into economic success for recent newcomers. Migrants are encountering serious and persistent challenges in finding employment commensurate with their qualifications and experience (Alboim, Finnie and Meng, 2005; Picot, Hou and Coulombe, 2007; Waslander, 2003). Recent migrants have higher unemployment rates and lower earnings than Canadian-born workers (Frenette and Morissette, 2003). Moreover, the gap between the average earnings of migrants and the average earnings of Canadian-born workers persists longer than in the past (Picot, Hou and Coulombe, 2007). While economic challenges are anticipated for refugees and for migrants who are sponsored by family members, the factors contributing to the persistent economic difficulties experienced by many skilled migrants are not well understood (Hiebert, 2002).[2]

Identifying the causes of migrants' persistent economic difficulties is particularly important for migrant women, who face even greater challenges in the Canadian economy than their male counterparts. Confirming a pattern that was identified in 1996, migrant women had higher unemployment, lower earnings and lower labour force participation rates than those of equally skilled Canadian-born women in 2001 (Tastsoglou and Preston, 2005), and they were almost 50 per cent more likely than migrant men to be underemployed (Galarneau and Morrissette, 2004). The Longitudinal Survey of Immigrants to Canada, (LSIC), a panel survey of those who arrived in Canada in 2000 and 2001, indicates that gender differences in every economic indicator begin within six months of arrival and the disparities between men's and women's economic success often increase during the first four years of settlement (Hawthorne, 2006).

Our study builds on previous findings by focusing on skilled migrant women. We compare the economic circumstances of skilled migrant women who differ in terms of their marital status, racial identities, fluency in Canada's official languages, and period of arrival, factors thought to influence women's decisions to enter the labour market and their success in obtaining remunerative and appropriate employment (Boyd, 1992; Chard, Badets and Howatson-Leo, 2000; Reitz, 1998). We examine labour force participation rates in detail, an approach that allows us to consider women who do not enter the paid labour market, a growing share of all migrant women (Hawthorne, 2006). The study concentrates on the experiences of employees.[3]

THE ECONOMIC EXPERIENCES OF FEMALE MIGRANTS

In the growing Canadian literature about the economic experiences of migrants, many studies still aggregate migrants or concentrate on migrant men (Alboim, Finnie and Meng, 2005; Aydemir, 2002, 2003; Aydemir and Skuterud, 2004; Schellenberg and Maheux 2007; Waslander, 2003). The lack of attention to gender is surprising since the economic experiences of migrant women (Chard, Badets and Howatson-Lee, 2000; Hawthorne, 2006; Hiebert, 2003; Preston and Giles, 1996; Reitz, 1998; Tsatsoglou and Preston, 2005) are even more disheartening than those reported for migrant men. Among full-time workers, the most privileged segment of the workforce, foreign-born women have lower earnings than foreign-born men and Canadian-born workers of both sexes. Migrant women are more likely to work in manual occupations than their Canadian-born counterparts, who are overrepresented in managerial, professional and clerical occupations (Tastsoglou and Preston, 2005). Migrant women are also more likely than their Canadian-born counterparts, and than Canadian-born and migrant men, to be unemployed.

In the last half of the 1990s, sustained economic growth caused the national unemployment rate to fall. Migrant men who arrived between 1991 and 2001 benefited substantially, with unemployment rates that declined faster than those of Canadian-born men. During the same time period, the disparity in unemployment rates between Canadian-born and foreign-born women persisted. In 2001, the unemployment rate for migrant women arriving in the preceding ten years was 14.4 per cent, more than double the unemployment rate of 6.1 per cent for Canadian-born women (Tastsoglou and Preston, 2005).

Examining unemployment rates in more detail underscores the precarious position of many migrant women in the Canadian labour market. The unemployment rate for migrant women who arrived in the 1990s is 15.7 per cent, almost three times higher than the unemployment rate for all migrant women (Table 9.1). Visible minority migrant women, who now account for the majority of recently arrived migrant women, also have unemployment rates higher than those for Canadian-born women from the same visible minority group. For example, the unemployment rate for migrant women who identify as West Asian and Arab is 16.6 per cent, whereas it is only 5.4 per cent for Canadian-born women from this visible minority group. Only the unemployment rates for Filipinas are lower for migrants than for the Canadian-born and the disparity in unemployment rates is small: 5.5 per cent for Canadian-born women of Filipino background and 4.6 per cent for migrant women from the same background. The reversal may be due, in part, to the predominance of Filipinas in the live-in caregiver programme that requires an offer of employment prior to arrival in Canada.

Marital status also influences unemployment rates. Married women have lower unemployment rates than single, separated, divorced or married women of working age, regardless of their birthplaces. With an unemployment rate of 5.0 per cent, married Canadian-born women are the least likely of all Canadian-born women to be unemployed. Married migrant women also reported a low unemployment rate when compared with single, separated, divorced and widowed migrant women: 7.5 per cent versus 7.7 per cent for all migrant women and a high of 10.4 per cent for migrant women who are separated.

High unemployment rates are accompanied by the withdrawal of sizable numbers of migrant women from the paid labour force. In 2001, 82.4 per cent of Canadian-born women between the ages of 25 and 44 years were in the paid labour force, compared with only 75 per cent of all migrant women in the same age group (Tastsoglou and Preston, 2005). For recent migrants from this age group, only 65 per cent of women were in the labour force. Less active in the paid labour market than Canadian-born women, migrant women are also less likely to participate in the labour market than migrant men.

The causes of migrant women's recent economic difficulties are not well understood. Recently arrived migrant women, those who settled in Canada between 1991 and 2001, are better educated than earlier cohorts of migrant women. Approximately one-third, 33.1 per cent, have at least one university degree, a high percentage compared with the 22 per cent of migrant women who arrived between 1971 and 1990 that are equally well educated

Table 9.1 Unemployment Rates of Canadian-Born and Migrant Women by Period of Arrival, 2001

| | Canadian-Born | All Migrants | Migrants by Period of Arrival | | | | | |
			Before 1961	1961–1970	1971–1980	1981–1990	1991–1996	1996–2001
Unemployment Rate	4.6%	5.8%	3.6%	4.2%	5.1%	6.7%	9.0%	15.7%
N	4,284,200	1,120,820	66,750	126,715	268,480	278,125	202,190	178,555

Source: Gender and Work Database, MIGCNS 1-2.

(Statistics Canada, 2003). Education does not always improve migrant women's economic prospects in the ways that we expect. Among women who arrived in Canada in the 1990s (Badets and Howatson-Leo, 1999; Tastsoglou and Preston, 2005), foreign-born women with at least one university degree are less likely to participate in the paid labour force than their Canadian-born counterparts. In comparison, the labour force participation rates for migrant women and Canadian-born women with less than a high school education are very similar. Education does enhance the earnings of migrant men and women. Recent migrants with more education reported higher earnings in 2001 than their less educated counterparts, although the earnings of foreign-born workers were not as high as those of equally well-educated Canadian-born workers (Chiu and Zietsma, 2003).

Several hypotheses have been proposed to explain the economic difficulties currently faced by migrant women (Boyd, 1992; Reitz, 1998). Labour market barriers undoubtedly contribute to their dismal economic circumstances. Faced with employers' unwillingness to recognise the value of foreign education and work experience abroad and lengthy and expensive accreditation processes for many regulated professions (Brouwer, 1999; Lopes, 2004), migrant women often settle for poorly paid and precarious jobs for which many are overqualified. As a result, they are more likely to be unemployed and to have low earnings than other workers in the Canadian labour market. The presence of labour market barriers means that the disparities in unemployment rates, occupational attainments, and earnings are likely to be greatest for the most recently arrived women, who are most vulnerable to devaluation of their foreign credentials and experience.

Migrant women may also withdraw from the labour market rather than accept jobs for which they are overqualified. Withdrawal may be a temporary situation as migrant women who are married stay at home to settle their families after arriving in Canada (Chard, Badets and Howatson-Leo, 2000). If this is the case, the gap in labour force participation should be higher for women who are married and should also diminish with length of residence in Canada. As settlement proceeds, we expect that migrant women will return to the paid labour market.

Limited fluency in Canada's official languages may also reduce skilled migrant women's employment opportunities, causing them to withdraw from the labour market (Boyd, 1992; Hawthorne, 2006; Preston, 2004). According to this hypothesis, fluency in one of Canada's official languages is linked directly to migrant women's economic success. Women who report fluency in either English or French are more likely to enter the paid labour force successfully, finding work that pays earnings commensurate with their qualifications and experience (Chiswick and Miller, 2002; Hawthorne, 2006; Pendakur and Pendakur, 2002). Their counterparts who lack fluency in one of the official languages are less likely to be in the paid labour force, and, once they enter the workforce, they face far more precarious employment prospects.

Recently arrived migrant women are also more likely to be visible minorities who often encounter discrimination in Canadian labour markets (Cranford and Vosko, 2006; Kunz et al., 2000; Pendakur, 2000; Reitz, 1998). Since 1990, the majority of people admitted to Canada have arrived from Asia and the Pacific, Africa, and Central and South America (Citizenship and Immigration, 2006). Hiebert (1999, 2000) has argued that gender and visible minority status create a 'triple jeopardy' such that visible minority migrant women are paid less than men of European background born in Canada. According to this hypothesis, migrant women of colour are the most likely group of workers to be in low-wage precarious employment. As a result, there will be larger gaps in unemployment rates, earnings, occupations, and participation rates for migrant women who are visible minorities than for those from European backgrounds.

Although the hypotheses specify the separate effects of each social characteristic, the effects are likely to interact. The detailed information about individual migrant women needed to specify the strength and nature of these interactions is not available. As a result, we analyse the separate effects of each selected social characteristic.

SKILL, LABOUR FORCE PARTICIPATION AND OCCUPATIONAL ATTAINMENT

In 2001, the educational attainments of migrant women are highly bifurcated, with disproportionate numbers of well-educated and uneducated women (Table 9.2). Consistent with current policies that emphasize the value of skilled migrants, a higher proportion of migrant women than Canadian-born women have some university education. At the other end of the educational spectrum, the percentage of migrant women with less than a grade nine education is about double the percentage of Canadian-born women, 7.4% versus 3.2%. Migrant women who settled recently in Canada are also better educated than their predecessors. Among the women who settled in Canada in the 1990s, almost half, 46.9%, have some university education compared with 35.0% in the preceding decade (Table 9.2).

Educational attainment increases migrant women's participation in the paid labour market just as it increases Canadian-born women's participation (Table 9.3), although the participation gains from education are smaller for migrant women than for the Canadian-born. Almost nine out of ten Canadian-born women with at least one university degree are participating in the labour market, approximately double the participation rate of 44.4% for women with less than a Grade 9 education. Among migrant women with at least one university degree, the labour force participation rate is 80.4 per cent, less than 50 per cent higher than the rate of 55.9 per cent for migrant women with less than a Grade 9 education.

Table 9.2 Educational Attainment of Canadian-Born and Migrant Women, 2001

Level of Education	Canadian-Born	All Migrants	Migrants by Period of Arrival				
			Before 1961	1961–1970	1971–1980	1981–1990	1991–2001
Less than grade 9	3.2%	7.4%	4.9%	9.8%	7.7%	8.1%	6.5%
Grades 9–13 without high school graduation certificate	14.7%	12.4%	14.4%	13.9%	12.2%	13.1%	11.2%
Grades 9–13 with high school graduation certificate	16.3%	13.7%	19.9%	16.2%	13.0%	13.1%	12.8%
Trades certificate or diploma	2.9%	2.4%	3.2%	2.7%	2.6%	2.5%	2.1%
College without certificate or diploma	7.0%	6.0%	6.2%	6.5%	6.6%	6.6%	5.2%
College with certificate or diploma	25.5%	19.3%	20.1%	21.5%	22.5%	21.5%	15.4%
University without bachelor's degree or higher	10.9%	12.8%	11.4%	10.7%	12.8%	12.7%	13.8%
University with bachelor's degree or higher	19.5%	25.9%	19.9%	18.7%	22.4%	22.3%	33.1%
N (total)	4,284,200	1,120,820	66,690	126,570	268,285	277,940	380,640

Source: Gender and Work Database, MIG CNS I-2.

Table 9.3 Labour Force Participation Rates of Canadian-Born and Migrant Women by Educational Attainment and Period of Arrival, 2001

Level of Education	Canadian-Born	All Migrants	Migrants by Period of Arrival					
			Before 1961	1961–1970	1971–1980	1981–1990	1991–2001
Less than grade 9	44.4%	55.9%	56.5%	54.4%	61.8%	60.4%	48.7%
Grades 9–13 without high school graduation certificate	67.8%	66.5%	68.7%	71.3%	72.1%	69.8%	58.6%
Grades 9–13 with high school graduation certificate	78.6%	71.1%	78.6%	79.4%	78.1%	73.5%	60.6%
Trades certificate or diploma	81.9%	77.6%	78.4%	82.3%	82.3%	80.2%	70.2%
College without certificate or diploma	79.2%	75.4%	80.2%	81.6%	81.0%	77.6%	66.3%
College with certificate or diploma	86.0%	82.0%	84.4%	84.8%	85.9%	84.8%	74.4%
University without bachelor's degree or higher	85.8%	77.7%	82.2%	85.2%	83.9%	82.2%	69.5%
University with bachelor's degree or higher	89.7%	80.4%	88.6%	87.3%	88.3%	85.3%	73.3%

Source: Gender and Work Database, MIG CNS I-2.

Labour force participation is much lower for migrant women who arrived in the 1990s than for any other cohort. At each level of education, women who arrived in Canada in the past decade have lower rates of participation than earlier arrivals (Table 9.3). The disparities are substantial, varying between 12.9 per cent and 10.0 per cent when women who arrived in the 1980s and 1990s are compared. The incremental changes from one decade to another are much smaller for those who arrived prior to 1991 than for those who arrived during the 1990s. Either the first ten years of settlement are more difficult or migrant women who arrived in the 1990s encountered unusually challenging barriers in the labour market.

Country of birth is also associated with marked difference in labour force participation rates (Table 9.4). Comparison of the labour force participation rates for women from countries that are sources of female migrants with high rates of university education reveals that only women from selected countries such as the United States, Romania, and the Philippines have high participation rates. Women from the United States have high rates of labour force participation consistent with employers' recognition of their educational attainments, qualifications, and work experience and their fluency in English (Boyd, 1992; Reitz, 1998). Women from the Philippines and Romania are almost as likely as American women to participate in the paid labour force. The impact of immigration policy is evident in the high rates of labour force participation reported by Filipinas. Many Filipinas immigrated to Canada as live-in caregivers, with offers of Canadian employment that ensured immediate participation in the labour force (Bakan and Stasiulis, 1997). The participation rates for Korean and Iranian women are more typical of recent arrivals. In both instances, approximately two-thirds of women with some university education engage in paid work in Canada.[4]

Table 9.4 Labour Force Participation Rates for Migrant Women from Selected Countries of Birth, 2001

Country of Birth	An Undergraduate Degree or Higher			Some University		
	All	1981–1990	1991–2001	All	1981–1990	1991–2001
Iran	73.6%	85.0%	67.9%	66.3%	80.9%	57.7%
Korea	62.6%	76.9%	50.5%	61.4%	70.1%	53.4%
Philippines	86.6%	88.6%	84.7%	84.9%	86.5%	83.5%
Romania	86.2%	92.1%	85.1%	87.3%	90.6%	86.0%
United States	84.6%	84.4%	75.6%	79.1%	82.5%	67.3%

Source: Gender and Work Database, MIG CNS I-2.

OCCUPATIONAL ATTAINMENT

One explanation for skilled migrant women's low participation in the labour market emphasizes labour market barriers. Employers undervalue foreign educational qualifications, and the processes to gain recognition of foreign professional credentials are lengthy, costly, and complex (Alboim, Finne and Meng, 2005; Reitz, 2002). Newcomers also lack knowledge of Canadian job markets, and their social networks provide little information about job opportunities or referrals for jobs commensurate with their qualifications and experience (Preston and Man, 1999). Faced with these barriers, we expect that migrant women will not have the same occupational achievements as equally well-educated women who are Canadian-born (Table 9.5). Looking at migrant women with at least one university degree, the majority have professional occupations, but a smaller percentage holds these desirable jobs than the percentage of equally skilled Canadian-born women. The disparity is more than 13 per cent, 52.6 per cent for migrant women as against 66.2 per cent for Canadian-born women. Migrant women are also more likely than their Canadian-born counterparts to be in less prestigious occupations such as manual work, sales and service, and clerical occupations. Almost 14 per cent of migrant women with at least one university degree work in sales and service compared with only 7.1 per cent of equally educated Canadian-born women. Educated migrant women are also more likely to be clerical and manual workers, 17.0 per cent and 3.9 per cent, respectively, than Canadian-born women.

Table 9.5 Occupational Attainments of Canadian-Born and Migrant Women with an Undergraduate Degree or Higher, 2001

Occupation	Canadian-Born	All Migrants	Migrants by Period of Arrival		
			Before 1991	1991–1995	1996–2001
Managers	12.6%	11.1%	13.1%	9.5%	7.8%
Professionals	66.2%	52.6%	57.1%	47.3%	45.3%
Supervisors	1.4%	1.4%	1.5%	1.7%	1.4%
Clerical	11.8%	17.0%	16.4%	18.8%	18.5%
Sales and Service	7.1%	13.9%	9.9%	17.0%	19.9%
Manual Workers	0.9%	3.9%	2.0%	5.8%	7.2%
N	919,650	325,945	173,745	56,520	80,100

Source: Gender and Work Database, MIG CNS O-2.

Recently arrived migrant women suffer the greatest decline in occupational status. By 2001, women who had arrived before 1991 were more likely to be employed as professionals than those who settled in Canada in the 1990s (Table 9.5). Among those who arrived in Canada between 1996 and 2001, only 45.3 per cent of women who had at least one university degree were working in professional occupations, whereas 57.1 per cent of their counterparts who arrived before 1991 had professional occupations.

There are marked variations in the occupational distributions of migrant women from different countries. Approximately one-third of Korean and Filipina women who have at least one university degree are employed in professional jobs, compared with close to two-thirds of equally educated migrant women from Romania and the United States (Table 9.6). Skilled Filipina and Korean women are much more likely to have sales and service and manual jobs than women from the other countries. There are similar trends across countries of birth among women who have attended university without completing a degree. Overall, these trends highlight the need for migrant women to have at least one university degree to compete successfully for professional jobs. Among women without a degree, the proportions working in professional occupations fall by almost one half.

Among women from the five selected sending countries, period of arrival also has discernible effects on their occupations. Women who arrived before 1991 are more likely to be employed as professionals than those who settled in Canada in the 1990s. The trends are clear if we compare the occupational distributions of women from the United States who settled in Canada before 1991, between 1991 and 1995, and between 1996 and 2001 with those of Filipinas who settled in Canada during the same time periods (Table 9.7). Almost two-thirds of the women from the United States who arrived prior to 1991 work in professional positions compared with only 57.8 per cent of their counterparts who arrived between 1996 and 2001. The same trend is apparent for Filipinas, but the disparity in participation rates is much larger between migrant cohorts. While 40.3 per cent of Filipinas who settled before 1991 have professional jobs, only 26.2 per cent of those who arrived between 1996 and 2001 are working in professional positions.

EXPLAINING MIGRANT WOMEN'S LABOUR FORCE PARTICIPATION

In 2001, the labour force participation of migrant women who have some university education is lower than that of equally educated Canadian-born women. The disparity in participation rates is partially due to period of arrival, since recent newcomers are less likely to be in the paid labour force than those who arrived prior to 1991. Educated migrants are also less likely than their Canadian-born counterparts to achieve the professional positions for which they are qualified by virtue of their educational attainments.

Table 9.6 Occupational Attainments of Women from Selected Countries of Birth, 2001

Country of Birth		Managers	Professionals	Supervisors	Clerical	Sales and Service	Manual Workers	N
Women with an undergraduate degree or higher	Iran	10.5%	50.6%	1.1%	15.2%	19.9%	2.7%	6,070
	Korea	27.5%	34.8%	1.1%	11.8%	21.7%	3.1%	5,805
	Philippines	5.5%	31.8%	2.0%	24.9%	29.0%	6.8%	32,185
	Romania	6.7%	68.0%	1.7%	12.6%	7.7%	3.2%	7,285
	United States	14.6%	64.4%	1.0%	12.0%	7.1%	0.8%	22,820
Women with some university	Iran	10.7%	31.2%	0.6%	19.5%	35.1%	3.0%	2,045
	Korea	31.7%	16.5%	2.4%	11.0%	34.7%	3.7%	2,790
	Philippines	4.1%	17.8%	2.3%	21.8%	41.1%	12.8%	19,230
	Romania	9.7%	39.6%	2.2%	17.9%	24.3%	6.3%	1,380
	United States	13.9%	30.9%	4.4%	27.2%	19.9%	3.7%	9,515

Source: Gender and Work Database, MIG CNS O-2.

Table 9.7 Occupational Attainments by Period of Arrival of Migrant Women from the Philippines and the United States Countries of Birth, 2001

Country of Birth	Period of Arrival	Managers	Professionals	Supervisors	Clerical	Sales and Service	Manual Workers	N
United States	Before 1991	13.7%	65.7%	1.0%	12.2%	6.6%	0.8%	24,240
	1991–1995	17.3%	60.6%	1.1%	11.0%	9.5%	0.5%	2,445
	1996–2001	18.9%	57.8%	1.1%	11.9%	8.9%	1.5%	2,640
Philippines	Before 1991	6.6%	40.3%	2.2%	26.0%	20.9%	4.0%	17,515
	1991–1995	6.0%	24.5%	1.8%	24.8%	33.2%	9.7%	10,060
	1996–2001	3.2%	26.3%	1.9%	23.3%	37.1%	8.3%	9,495

Women with an undergraduate degree or higher

Source: Gender and Work Database, MIG CNS O-2.

Finally, there are significant differences in participation rates and occupational attainment among skilled migrant women from different countries of birth. A series of cross-tabulations provides some insight into the causes of downward mobility, revealing the effects of marital status, fluency in English, and visible minority status on migrant women's labour force participation.

Marital Status

The effects of marital status are different from our expectations. Marital status has little impact on the labour force participation of Canadian-born women, at least among women aged 25 to 54 years of age (Table 9.8). The participation rate hovers around 80 per cent for single, married, separated, and divorced women. Widows are less likely to be in the labour market with a participation rate close to 70 per cent. In 2001, being married does not reduce rates of participation in the labour force substantially, unlike widowhood that is associated with a marked decline in labour force participation.

Marriage has more impact on the participation rates of migrant women than on the participation rates of Canadian-born women. Migrant women who are single or divorced participate in the paid labour force at approximately the same rate as their Canadian-born counterparts, between 80.9 per cent and 80.5 per cent. Participation rates for migrant women who are widowed, married, and separated are lower, ranging from 65.6 per cent for widows to 76.7 per cent for separated women. The disparity in participation rates between Canadian-born and migrant women is largest for married women. Only 72.8 per cent of married migrant women are in the paid labour force compared with 80.6 per cent of Canadian-born women who are married.

Disaggregating the migrant population by period of arrival reveals a more complex story. Only the labour force participation rates of recent arrivals are related to marital status. Among women who arrived in the 1990s, participation rates varied from approximately 80 per cent for single women to less than 60 per cent for widows. Married women had low participation rates of less than 65 per cent. The impact of marriage diminishes tremendously in earlier cohorts of migrants where the participation rates of married migrant women are much closer to the participation rates of married women who are Canadian-born (Table 9.8). Women who arrived in Canada between 1981 and 1990 report rates of labour force participation that vary relatively little with marital status. Although the data cannot tell us the reasons that marriage contributes to low participation rates among recent newcomers, the temporal pattern of the effects of marriage is consistent with the argument that married women withdraw from paid work in the initial settlement period to settle their families (Chard, Badets and Howatson-Leo, 2000). The trends also suggest that gender roles within the household may change with longer residence in Canada such that married women are more likely to enter the paid labour force.

Table 9.8 Labour Force Participation Rates of Women in Canada by Marital Status, 2001

	Canadian-Born	All Migrants	Migrants by Period of Arrival					
			Before 1961	1961–1970	1971–1980	1981–1990	1991–2001	
Never married (single)	82.0%	80.9%	80.8%	85.2%	85.6%	83.1%	79.2%	
Married (including common-law)	80.6%	72.8%	79.1%	78.1%	80.3%	77.7%	65.4%	
Separated	82.0%	76.7%	83.2%	83.1%	83.5%	77.1%	70.7%	
Divorced	82.3%	80.5%	82.9%	84.0%	83.2%	81.3%	75.6%	
Widowed	70.5%	65.6%	74.9%	69.2%	70.6%	67.2%	59.7%	
N (in labour force)	4,270,575	1,159,760	66,695	126,570	268,285	277,940	380,640	

Source: Gender and Work Database, MIG CNS K-1.

Country of birth also affects the association between marriage and participation in the labour force. Although marriage reduces participation in the labour force for all migrant women, the amount of the reduction in labour force participation varies (Table 9.9). The participation rate of married women from Korea is approximately 10 per cent lower than that of single women who emigrated from the same country. In contrast, married women who migrated from the Philippines and Romania have labour force participation rates that are only 5 per cent below those of single women from the same countries. The varied effects of marriage across countries of origin highlight the importance of migration history and the premigration context for each group of migrants.

The Effects of Language

Acquiring a fluent knowledge of Canada's official languages is a well-known challenge for newcomers (Lopes, 2004). Without fluent knowledge of English or French, well-educated migrants are often not able to obtain jobs for which they are otherwise qualified by education, experience, and training. Knowledge of English and French poses particular challenges for migrant women, who, historically, have had less knowledge of Canada's official languages than migrant men (Boyd, 1992).

Women with knowledge of both official languages have higher rates of labour force participation than those who know only one official language, who in turn have higher participation rates than migrant women who do not know either official language (Table 9.10). Knowledge of Canada's official languages increases the participation rates of migrant women, but it does not raise them equal to those of Canadian-born women. Migrant

Table 9.9 Labour Force Participation Rates by Marital Status for Migrant Women from Selected Countries of Birth, 2001

	Iran	Korea	Philippines	Romania	United States
Never married (single)	74.8%	76.5%	89.2%	88.1%	83.7%
Married (including common-law)	64.0%	63.2%	83.6%	82.8%	78.3%
Separated	61.9%	60.6%	86.7%	83.1%	79.4%
Divorced	77.9%	72.5%	89.0%	87.6%	85.0%
Widowed	57.7%	58.1%	86.1%	88.9%	72.8%
N (in labour force)	13,115	13,795	78,565	13,265	56,610

Source: Gender and Work Database, MIG CNS K-1.

Table 9.10 Labour Force Participation Rates of Women in Canada, by Knowledge of Official Languages, 2001

Language Knowledge	Canadian-Born	All Migrants	Migrants by Period of Arrival					
			Before 1961	1961–1970	1971–1980	1981–1990	1991–2001
English Only	80.8%	75.5%	79.5%	79.2%	81.5%	79.7%	69.3%
French Only	75.2%	65.0%	71.3%	63.6%	71.6%	69.4%	61.2%
Both English and French	85.1%	80.2%	82.3%	83.7%	84.7%	81.4%	75.5%
Neither English nor French	39.6%	48.3%	41.1%	44.4%	56.8%	58.6%	46.1%
N (in labour force)	4,270,575	1,159,765	66,695	126,565	268,285	277,940	380,635

Source: Gender and Work Database, MIG CNS J-1.

women who know one or both official languages still have lower rates of participation in the labour market than Canadian-born women with the same language skills.

Knowledge of official languages influences how migrant women's labour force participation changes with residence in Canada. Comparison of migrant women's labour force participation across period of arrival cohorts and groups, categorised on the basis of language knowledge, indicates that the effects of period of arrival are minor for women who know both official languages. Approximately three quarters of women who arrived in the 1990s and knew English and French are in the paid labour force, compared with more than 80 per cent of their counterparts who arrived earlier. The proportional increase in labour force participation rates is larger for women who know only one official language or for those who do not know an official language. For example, in 2001, the labour force participation rate for migrant women who knew only French increased from 61.2 per cent for those who arrived in the 1990s to approximately 70 per cent for those who arrived in earlier decades. Equally large increases in participation are evident after the first decade in Canada for migrant women who know only English and who know neither language.

The data do not allow us to examine directly the relationship between knowledge of official languages and education. However, if we look again at the experiences of women from the five countries with large numbers of well-educated women (Table 9.11), English is the best-known official language except for the Romanians. More than 40 per cent of Romanian women know both official languages. The extent to which women's language fluency improves with residency in Canada varies across the five countries of origin. The vast majority of women from the United States, the Philippines, and Romania are fluent in at least one of Canada's official languages, so the percentages who know an official language don't increase much with residence in Canada. Among Iranian and Korean women, recent newcomers who settled in the 1990s are more likely to know at least one official language than their counterparts who arrived in earlier decades.

Labour force participation is higher for migrant women who know both of Canada's official languages (Table 9.12). For example, the participation rate for women from the United States who know only English is 78.5 per cent compared with a participation rate of 84.0 per cent for women from the same country of birth who know both languages. Knowledge of both official languages is also associated with higher participation rates for women who have emigrated from Iran, Korea, and Romania. The situation of Filipinas is exceptional once again. For Filipinas, knowledge of English is crucial. Among those who arrived most recently, during the 1990s, the rate of labour force participation for those who know only English exceeds that for Filipinas who know both languages.

Table 9.11 Knowledge of Official Languages by Period of Arrival for Migrant Women from Selected Countries of Birth, 2001

		English Only	French Only	Both English and French	Neither English nor French	N
Iran	All	81.5%	1.0%	14.2%	3.3%	19,870
	1991–2001	78.6%	NA	20.3%	0.6%	5,675
	1981–1990	83.6%	1.2%	10.6%	4.6%	13,265
Korea	All	83.8%	NA	4.0%	11.9%	21,235
	1991–2001	89.2%	0.6%	6.9%	3.3%	4,270
	1981–1990	78.7%	0.4%	2.5%	18.4%	12,440
Philippines	All	97.2%	NA	2.5%	NA	92,155
	1991–2001	97.4%	NA	2.3%	0.2%	24,755
	1981–1990	97.6%	NA	2.1%	0.2%	48,705
Romania	All	52.9%	4.8%	41.6%	0.8%	15,880
	1991–2001	61.2%	5.3%	32.6%	0.9%	3,325
	1981–1990	50.7%	4.9%	43.5%	0.9%	11,410
United States	All	83.6%	0.7%	15.7%	NA	71,320
	1991–2001	86.1%	0.6%	13.3%	NA	16,100
	1981–1990	87.6%	0.3%	12.0%	NA	12,610

Source: Gender and Work Database, MIG CNS J-1.
NA—Percentages are not available; N is too small for valid estimates.

Visible Minority Status

Discrimination in the labour market reduces the occupational attainments of visible minorities (Pendakur, 2000), and it may discourage highly skilled women from entering the labour market (Preston and Man, 1999). There is a significant difference in labour force participation between Canadian-born women and migrant women who are visible minorities, 83.1 per cent versus 72.1 per cent (Table 9.13). The difference in participation rates is due mainly to the low participation rates of visible minority migrant women who arrived in the 1990s. Only about two-thirds, 65.6 per cent, of visible minority migrant women who arrived in the 1990s participate in the paid

Table 9.12 Labour Force Participation Rates by Knowledge of Official Languages of Migrant Women from Selected Countries of Birth and Period of Arrival, 2001

		English Only	French Only	Both English and French	Neither English nor French	N (in Labour Force)
Iran	All	66.27%	46.15%	73.94%	30.53%	13,115
	1991–2001	80.11%	NA	77.92%	0.00%	4,495
	1981–1990	60.14%	42.42%	68.33%	32.23%	7,890
Korea	All	67.27%	NA	71.18%	46.25%	13,795
	1991–2001	77.95%	100.00%	83.05%	50.00%	3,305
	1981–1990	56.74%	66.67%	46.03%	45.20%	6,760
Philippines	All	85.35%	NA	84.70%	NA	78,565
	1991–2001	87.18%	NA	84.48%	30.77%	21,520
	1981–1990	83.27%	100.00%	84.00%	57.89%	40,550
Romania	All	83.25%	75.66%	85.83%	28.00%	13,265
	1991–2001	84.28%	77.78%	89.40%	0.00%	2,825
	1981–1990	82.82%	74.77%	85.50%	30.00%	9,485
United States	All	78.53%	76.42%	83.98%	NA	56,610
	1991–2001	78.97%	84.21%	85.01%	NA	12,845
	1981–1990	70.71%	77.78%	72.61%	NA	8,955

Source: Gender and Work Database, MIG CNS J-1.
NA—Percentage are not available; N is too small for valid estimates.

labour market compared with 77.9 per cent of those who arrived a decade earlier.

The aggregate figures mask the diverse labour market participation of visible minority women. The majority of Canadian-born visible minority women have participation rates equal to or exceeding the participation rate of approximately 80 per cent for all Canadian-born women. The exceptions are Canadian-born Arab and West Asian women, whose participation rates of 65.5 per cent and 73.6 per cent, respectively, are low (Table 9.13). Participation rates for migrant women who are also visible minorities are even more varied than those for visible minority women born in Canada. For example, the participation rate of Filipina migrant women is 85.8 per cent, higher than the participation rates for all Canadian women or for all

Table 9.13 Labour Force Participation Rates by Visible Minority Status of Women in Canada, 2001

	Canadian-Born	All Migrants	Migrants by Period of Arrival	
			1981–1990	*1991–2001*
Total visible minority population	83.1%	72.1%	77.9%	65.6%
Chinese	86.5%	68.9%	76.2%	61.2%
South Asian	83.6%	71.7%	78.1%	64.7%
Black	80.1%	79.9%	82.4%	74.4%
Filipino	86.6%	85.8%	87.6%	83.5%
Latin American	82.1%	71.3%	76.4%	69.0%
Southeast Asian	82.5%	71.4%	74.3%	64.0%
Arab	73.6%	53.3%	60.8%	47.5%
West Asian	65.5%	60.3%	74.3%	55.1%
Korean	82.0%	59.2%	77.5%	54.4%
Japanese	86.0%	57.5%	66.4%	61.4%

Source: Gender and Work Database, MIG CNS I-1.

migrant women. However, among visible minority migrant women, Filipinas and blacks are the only groups whose participation rates come close to the average for all Canadian women. Although Arab and West Asian migrant women still have low participation rates of 53.3 per cent and 60.3 per cent, respectively, they are almost matched by the rates of 59.2 per cent and 57.5 per cent reported by Korean and Japanese women.

Participation rates for women from each of the visible minority groups increase with residence in Canada. In each visible minority group, the participation rates for women who settled in Canada in the 1980s are higher than for their counterparts who arrived in the 1990s. Nevertheless, it is difficult to account for the initial disparity in labour market participation by visible minority migrant women. Recently arrived migrant women who are visible minorities are well educated. Among those 25 to 44 years of age[5] who arrived in the 1990s, almost half, 47.7 per cent, had some university education. This is a substantial increase over the 34.5 per cent of visible minority migrant women who arrived in the 1980s with some university education. On the surface, at least, current trends in labour market participation of many visible minority migrant women, specifically those who are not Filipinas, are consistent with the hypothesis that visible minority migrants are discouraged from entering the labour market by discrimination (Pendakur, 2000). Careful examination of the interacting effects of various social characteristics and the work histories and job search experiences of migrant women is needed to evaluate this hypothesis fully.

CONCLUSION

Well-educated migrant women are still not participating in the paid labour market at the same rates as their Canadian-born counterparts. The trends identified initially in 1996 (Chard, Badets and Howatson-Leo, 2000) have persisted despite the buoyancy of the Canadian economy between 1996 and 2001. Although the disparity in participation rates declines with residence in Canada, it does not disappear. Moreover, educated migrant women are also less likely than Canadian-born women with equivalent education to achieve the professional positions for which they are qualified. There is also support for each of the hypotheses proposed to explain the low participation rates of well-educated migrant women. Marriage is associated with lower participation rates but only for migrant women who arrived in the 1990s. Longitudinal analysis is needed to ascertain whether the women who arrived in the 1990s are now moving into paid employment in larger numbers than in 2001. Fluency in one of Canada's official languages increases labour force participation rates for all migrant women. However, among recent migrant women, the majority of whom possess knowledge of at least one of Canada's official languages, knowledge of an official language does not increase participation rates to the levels reported by earlier arrivals. There is compelling evidence that visible minority status reduces migrant women's participation in the labour force. Among the Canadian-born, visible minority status is associated with high levels of labour force participation; however, among migrant women, participation rates are low for women from most visible minority groups.

The findings highlight the diversity of migrant women's experiences in Canadian labour markets. The analysis of labour force participation rates confirms previous suggestions (Hiebert, 1999, 2000; Pendakur, 2000) that the labour market experiences of migrant women from visible minority groups are often quite different from those of Canadian-born women from the same visible minority backgrounds. For example, among the Canadian-born, only Arab and West Asian women have low rates of participation in the paid labour market. Among migrant women, Filipina and black migrant women are exceptional because of above-average rates of participation, unlike migrant women from every other visible minority group, whose participation rates are well below the average for Canadian-born women. Sophisticated analysis that explores the diverse experiences of women from different visible minority groups will aid efforts to understand the effects of marital status and language fluency that also vary across visible minority groups.

The diverse experiences of migrant women who arrived in the 1980s and those who arrived in the 1990s are starkly apparent from our analysis. Several recent studies from Statistics Canada have also emphasized differences in the labour market experiences of these two cohorts (Aydemir and Skuterud, 2004; Frenette and Morisette, 2003). Yet, we still do not know whether length of residence itself alters migrant women's experiences in the labour market or whether, and more likely, the circumstances that migrants encounter upon arrival influence how length of residence in Canada affects labour market experiences (Aydemir, 2003). Cross-sectional analysis that is inherently partial cannot answer this question. To document the strategies by which foreign-born women overcome the challenges to successful economic integration, detailed information about individual women's experiences is needed.

Our analysis of census information has identified troubling trends in migrant women's economic integration in Canada. The downward mobility of many educated migrant women suggests that current Canadian immigration policies that favour well-educated migrants do not ensure their economic success. The findings raise profound questions about the reception of migrants in contemporary Canadian society. As Reitz (1998, 2002) has suggested, researchers and policymakers would do well to consider the roles of all types of institutions in facilitating the validation and utilization of the human capital that qualifies so many newcomers for admission to Canada and that, currently, they are unable to utilize fully in Canada. Our analysis confirms that improving economic prospects of migrant women requires moving beyond the traditional focus on the labour market to consider the specific gendered challenges confronting many migrant women (Kofman and Raghuram, 2006). In a period of growing concern about work-life balance, attention to the challenges facing migrant women is likely to benefit all Canadian women struggling to shoulder the competing expectations of paid and unpaid work.[6]

NOTES

1. Migrants refers only to people admitted as permanent residents. Asylum seekers and temporary migrants are not considered in this chapter.
2. While skilled migrants are doing better economically than either refugee- or family-sponsored newcomers, all economic indicators show that the economic circumstances of skilled migrants lag those of their Canadian-born counterparts with equivalent educational and linguistic qualifications and work experience (Hawthorne, 2006).
3. Despite the sometimes 'heroic' descriptions of the self-employed, migrant men and women who are self-employed have low incomes, often lower than those of employed migrants (Li, 1999).

4. Participation rates are similar in the three metropolitan areas that are the main destinations for migrants in Canada. Reflecting its dominance as the single largest destination for migrants and the largest urban economy in Canada, the participation rates for Canadian-born and migrant women in the Toronto metropolitan area are almost identical to the national rates. In Vancouver, participation rates are slightly lower than the national average, while in Montreal, participation rates for less educated migrant women are higher than the national rates. Overall, the labour force participation rates for migrant women are similar across the metropolitan areas, so the remainder of our analysis concentrates on national trends.

5. Data about education and visible minority status are not reported for 25–54 years of age.

6. We acknowledge with gratitude access to the Gender and Work Database funded by the Social Sciences and Humanities Research Council and Statistics Canada. This chapter is based on an earlier working paper undertaken with Wenona Giles that was funded by York University and is available at http://www.genderwork.ca/conference/Preston-Giles_edited_final.pdf.

Part III

Structural Discrimination and Strategies of Response

10 Barriers to the Labour Market
Refugees in Britain

Alice Bloch

Research evidence repeatedly demonstrates that some refugees arrive in the UK with high levels of skills, qualifications and premigration work experience and that there is diversity of skills, qualifications and work histories between and within different refugee groups (Charlaff et al., 2004; Kirk, 2004). However, rates of employment are very low among refugees in the UK and even the most highly skilled refugees, who have found employment, are for the most part in jobs that are not commensurate with their skills, qualifications and premigration labour market experiences (see, for example, Dumper, 2002; Charliff et al., 2004; Stewart, 2003; Employability Forum, 2006).

This chapter draws on research carried out for the Department for Work and Pensions of 400 refugees,[1] with permission to work, to explore their labour market position and the barriers they face to labour market participation and/or obtaining employment commensurate with their skills and qualifications (Bloch, 2002). Somali, Tamil, Kosovan, Turkish and Iraqi refugees, including Kurdish refugees, with permission to work were interviewed using face-to-face interview methods and translated questionnaires. Interviews were carried out in London, the North West, North East, Midlands and Yorkshire and Humberside between February and May 2002. Quotas were set for the key explanatory variables of country of origin, gender, age, length of residence in the UK and region.[2] In the final sample, eighty interviews were carried out with respondents from each of the five countries of origin of which half were with men and half were with women. Three-quarters of respondents had been less than thirty-five years old on arrival in Britain while 56 per cent were less than thirty-five years or older at the time of the survey. Just under half (46 per cent) had been in Britain for less than three years while the rest had been in Britain for three years or more. More than half the interviews were carried out in London (54 per cent) while the remaining interviews were carried out in the other four regions included in the study and took place for the most part in the urban centres of Manchester, Birmingham, Newcastle, Leeds and Sheffield.

The research findings were used to help inform the Department for Work and Pensions refugee employment strategy (DWP, 2005) and also contributed to the Home Office strategy detailed in *Integration Matters: A National Strategy for Refugee Integration* (Home Office, 2005). Employment is seen as a crucial indictor of integration, which the Home Office (2005) maintains takes place when refugees achieve their full potential as members of British society, contribute to the community and are able to access the services to which they are entitled. In terms of achieving full potential, the Home Office identifies two factors as being crucial, 'the ability to communicate effectively in English and gaining employment appropriate to their skills and ability' (2005: 20). Therefore, identifying and addressing the barriers that refugees face in accessing the labour market and using their skills has been an important component of government thinking around integration.

This chapter will first describe the skills base of refugees in terms of pre-arrival skills, qualifications and employment experience and will then examine capacity building and the acquisition of new skills in the UK. With this skills base in mind, the chapter will focus on the employment profiles of refugees, disaggregating by demographic and human-capital variables. The chapter will then examine barriers to the labour market, successful routes to employment and job-seeking methods among refugees. It will highlight the impact of social networks and community links for refugees as well as the extent to which dispersal and the placement of refugees by local authorities in other localities impact on employment prospects and job-seeking strategies.

THE SKILLS BASE OF REFUGEES

Refugees arrive in the UK with different premigration educational, employment and skills bases and some acquire new skills and qualifications in the UK. A skills audit of Zimbabwean refugees carried out by the Home Office (Kirk, 2004) highlights the diversity of educational and employment backgrounds among refugees both between groups and within groups. In the Home Office skills audit, the majority of Zimbabweans—92 per cent—arrived in the UK with a qualification and there was little difference between men and women. In contrast, there was a large gender disparity among Somali men and women. Fourteen per cent of Somalis arrived in the UK with a qualification, and more than half (55 per cent) of Somali refugee women had received no formal education compared with a quarter (24 per cent) of Somali men.

An analysis of the occupations of refugees prior to migration in the Home Office skills audit compared with data from the Labour Force Survey found that refugees were, on average, in higher skilled employment prior to migration than the UK population. Twenty-two per cent of refugees had been in managerial and senior positions compared with 15 per cent of the UK population, 15 per cent of refugees had been in professional occupations compared with 12 per cent of the UK population and 23 per cent had been in skilled trades compared with 12 per cent of the UK population (Kirk, 2004; Begum, 2004). However, once in the UK, as this chapter will show, such skills and experiences are not transferred. Instead, the minority of refugees who are working are either underemployed or were working in secondary sector jobs.

One important link that has been consistently emphasised in research is the importance of English language as a key component in the successful economic integration of refugees (Bloch, 2002; Home Office, 2005; Phillimore and Goodson, 2006). In the DWP study, 6 per cent of respondents spoke English fluently and 12 per cent spoke English fairly well on arrival to the UK while half of all respondents had no English language. There was a correlation between country of origin and English on arrival based on former colonial or protectorate links with the UK. As a consequence of these former links, refugees from Sri Lanka, Somalia and Iraq were more likely to speak English than those from Turkey or Kosova. However, by the time of the survey, a marked improvement in English was apparent with only 7 percent without any English language, more than a fifth could speak English fluently (21 per cent) and 39 per cent spoke English fairly well. Table 10.1 show the differences in spoken English between and within country of origin groupings. Refugees from Kosova and Turkey were the least proficient, especially women, while Somali men and Iraqi women were the most likely to have high levels of English-language proficiency.[3]

In addition to arriving in the UK with a diverse range of English-language competencies, refugees also arrive with different skills sets and qualifications. In this study, more than half (53 per cent) had a qualification on arrival to the UK that was higher than the 40 per cent found in the Home Office skills audit (Kirk, 2004). Sixteen per cent arrived with a degree, higher degree or professional qualification—which was nearly one-third (30 per cent) of all those with qualifications. There were differences between refugees from different countries in terms of having a qualification and the level of qualifications. Figure 10.1 shows that refugees from Sri Lanka were most likely to have a qualification (81 per cent), followed by refugees from Iraq (64 per cent), Turkey, (59 per cent), Kosova (40 per cent) and Somalia (21 per cent). Refugees from Iraq were the most likely to have a degree, a postgraduate qualification or professional qualification.

Since living in the UK, a fifth of those interviewed had obtained a qualification in the UK of which a fifth were either postgraduate, degree or professional qualifications. Most of those with higher level qualifications from the UK were from either Somalia or Sri Lanka.

Finally, refugees arrive with valuable labour market experience that could be used in the UK. Before arrival, 42 per cent had been working while an additional 8 per cent had been employed at some time. Some had professional jobs that could be carried out in the UK (20 respondents had been teachers); skilled trades such as carpenters and plumbers (17 had been in these professions) and office and administration (20). There were also 13

Table 10.1 Proficiency in Spoken English at the Time of the Survey by Country of Origin and Sex

| | Somalia | | Turkey | | Iraq | | Sri Lanka | | Kosova | | |
	Men	Women	Men	Women	Men	Women	Men	Women	Men	Women	%
Fluently/fairly well	33	21	24	13	21	34	27	30	23	13	60
Slightly/not at all	7	19	16	27	19	6	13	10	17	27	40
Total number	40	40	40	40	40	40	40	40	40	40	400

Frequencies

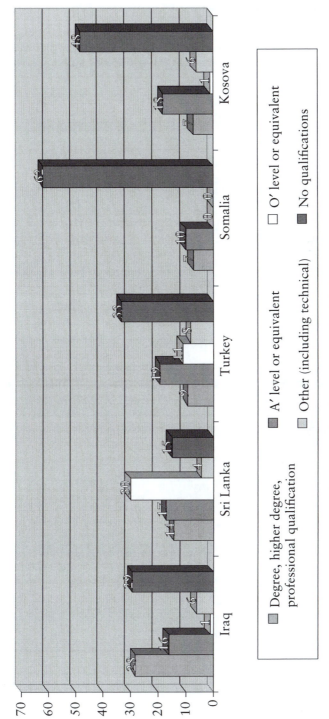

Figure 10.1 Qualification by country of origin, frequencies. Base number: 399. Missing: 1.

health-care professionals including doctors, nurses and dentists. Others had jobs that were more country or context specific such as farming (14 had been farmers). Having detailed the skills base of refugees, the next section will consider the position of refugees in the UK labour market and shows an underutilisation of these skills.

THE POSITION OF REFUGEES IN THE LABOUR MARKET

Research with refugees has consistently shown low levels of employment and among the minority of refugees who are working underemployment (see, for example, Dumper, 2002; Charliff et al., 2004; Phillimore and Goodson, 2006). In this study, 29 per cent of respondents were working at the time of the survey and a further 19 per cent had worked in the past but were no longer working. Table 10.2 shows the key variables that interact with labour market participation. There was a clear relationship between level of English-language fluency and likelihood of employment as well as other human-capital variables such as training and qualifications.

Table 10.2 Proportion of Refugees Who Were Working by Key Explanatory Variables

Labour Market Participation	*Working %*	*Total No.*
[1]Spoken English		
Fluently	52	83
Fairly well	31	156
Slightly	14	133
Not at all	11	28
Whether arrived with a qualification		
Yes	37	212
No	18	187
Whether obtained qualification in Britain		
Yes	51	81
No	23	319
Length of residence		
Less than three years	20	185
Three years or more	36	215

Table 10.2 Proportion of Refugees Who Were Working by Key
Explanatory Variables (*continued*)

Labour Market Participation	Working %	Total No.
Immigration status		
'Secure status': Refugee, ILR, Naturalised British/ EU citizen	31	229
ELR	35	82
Temporary admission	25	89
Participation in training		
Yes: past	67	30
Yes: current	41	17
No	25	353
Region		
North East	17	18
North West	19	32
Midlands	26	76
London	30	216
Yorkshire and Humberside	35	58
Country of origin		
Somalia	20	80
Turkey	31	80
Iraq	24	80
Sri Lanka	45	80
Kosova	22.5	80
Age		
Less than 35	36	221
35 or over	20	179
Sex		
Men	41.5	200
Women	15.5	200
Total number	114	400

[1]English-language standard was self-assessed.

Men were much more likely to be working than women (42 per cent and 16 per cent, respectively), and there were differences by country of origin, with refugees from Sri Lanka more likely to be working than others. Country origin also interacted with gender and employment; Tamil men are the group most likely to be working (36 out of 40) while Iraqi women and Kosovan women are the least likely to have a job (1 out of 40 and 3 out of 40, respectively). The only community for which there is little difference is the labour market activity between men and women is the Somali community; in the other four groups men were more likely to be working than women. The data in Table 10.2 also show a regional variation, with higher levels of refugee employment in Yorkshire and Humberside and London than elsewhere. This is due in part to the demand for service-sector employees in these areas—jobs that are unskilled and low paid. Research with ethnic minorities also shows the regional and local variations in labour market participation (Ho and Henderson, 1999; Commission for Racial Equality, 2006).

Table 10.2 also shows the link between English-language proficiency and employment, with more than half of those fluent in English employed at the time of the survey. An analysis of people's experiences of using English in everyday contexts found that while those who spoke English fluently were able to use English for job seeking without problems, there were some problems using English within this context among those who spoke English fluently, with a quarter stating that they found it difficult. Part of the difficulties of using English for job seeking could relate to the technical language used in some areas of application and types of employment. Other research has also explored aspects of English language within the job search context and found that accent has an affect on employment opportunities as does the interaction of accent with ethnicity which McKay et al. (2006) argue raises questions surrounding employer discrimination. Research with prospective employers has reinforced the emphasis placed on English and was given as a reason for rejecting refugee applicants (Hurstfield et al., 2004).

While having qualifications on arrival to the UK does affect labour market participation, the correlation between employment and qualifications gained in the UK is stronger, as Table 10.2 shows. This is due in part to the greater propensity of those with a UK qualification to have higher levels of English-language proficiency than others. It is also because some refugees and other migrants who arrive in the UK with qualifications can find that their qualifications are not recognised, which means that professionals are unable to practice their professions without retraining. One case that is frequently highlighted is the case of health-care professions, including doctors (Dumper, 2002; Stewart, 2003).

Moreover, refugees with premigration qualifications rarely attempt to have their qualifications recognised or transferred to their UK equivalents. In the DWP study only 15 per cent of those with qualifications had tried to have them transferred. The reasons for not trying to get qualifications recognised varied, but mentioned most often were: lack of English (16 per cent), not having certificates, probably due to the circumstances of exile which might result in them leaving home under acute circumstances (16 per cent), didn't need to get them recognised (16 per cent), didn't know how to (12 per cent), no time or too busy to do so, including with family (10 per cent) and didn't think to (7 per cent). The fact that 16 per cent did not have their certificates with them reflects some of the problems faced by refugees when trying to gain access to education and employment commensurate with skills and experience, namely, the lack of evidence of past achievements. A combination of UK structural factors to the recognition of qualifications, circumstances of flight and exile and a lack of information can all add to barriers to the labour market and the use of premigration skills and qualifications.

One of the consequences of these barriers is that those refugees who were working at the time of the survey were concentrated in a limited number of jobs and sectors of employment. Sixty per cent were working in just four areas: catering; interpretation and translation; shop and cashier work and administration and clerical work. This is in contrast to the diversity of employment and occupations prior to migration that included professional jobs such as teachers and engineers and skilled trades like builders, mechanics and carpenters. Many of the jobs that refugees had had before coming to the UK were replicable in the UK context, though some, such as farming, were not. Other research has also highlighted the range of occupations among refugees prior to migration and the greater likelihood of them having been in professional and skilled employment prior to migration than in the UK (Dumper, 2002; McKay et al., 2006; Bloch, 2008).

Refugees who were working in the UK had for the most part been in their current job for a relatively short period of time. Around a quarter (26 per cent) had been in their job for less than six months and two-thirds (65 per cent) for less than a year and a half. There was little transition or progression between jobs—nearly two-thirds (60 per cent) had been unemployed before finding their current job. Those who were working were for the most part clustered in low-paid work with poor terms and conditions of employment. On average, refugees were earning only 79 per cent of that earned by ethnic minorities as a whole and 11 per cent were earning less than the minimum wage. Employment was insecure, with a quarter in temporary jobs; a third were working part time; nearly half (47 per cent) were working unsocial hours; and 42 per cent worked a minimum of two weekends a month. Less than half were entitled to holiday pay and only a third had been offered any on-the-job training, reflecting the sometimes casual nature of refugee employment. In sum, refugees as an aggregated group find themselves more disadvantaged than others in the labour market. However, there are variations between and within refugee groups in terms of labour market experiences and barriers to employment, which means the policy interventions need to better recognise diversity (Morrice, 2007).

BARRIERS TO THE LABOUR MARKET

Barriers to the labour market, experienced by refugees, can be delineated into those that emerge from policy and other external factors, like discrimination and employer reluctance to check documentation, alongside the needs for individuals to build their capacity through the acquisition of new knowledge and skills (Green, 2005). In addition, refugees can also experience barriers that are a consequence of being a refugee such as trauma, stress and health problems as a result of torture or separation from family members, and like others from minority ethnic groups, refugees can and do experience racism and discrimination by prospective employers, which can include accent, religion and religious identity and nationality (Cabinet Office, 2003; Archer et al., 2005; McKay et al., 2006). Thus, the barriers faced by refugees involve the complex interaction of a number of factors.

The focus of the DWP research was first on the personal and human-capacity factors that interact with refugee employment, secondly on structural and policy barriers and thirdly on the social and cultural-capital factors that impact on access to the labour market, including social networks and the cultural aspects of job seeking. Table 10.2 shows the interaction of personal and human-capacity factors and the propensity for refugees to be working. For example, those with higher levels of English-language proficiency arrived in the UK with a qualification and/or had obtained a qualification in the UK, had participated in training, had been resident in the UK for three years or longer and were less than thirty-five years of age were most likely to be working.

Variations were also evident by region, with those in areas with more buoyant local economies, particularly in the low-paid unskilled service sector, more likely to find work. However, such trends should not hide the diversity among refugees. Some refugees are fluent in English, have high levels of qualifications (degree and higher) and professional work experience but still remain unemployed or underemployed while others arrive in the UK with limited or no English, without qualifications and work experience but are able to find employment.

When interviewees were asked what were the barriers they experienced to the labour market, a greater proportion identified personal and human-capacity factors before structural or employee barriers, though these were also in evidence, as Table 10.3 shows. English language and lack of work experience in the UK were clearly viewed as the most immediate barriers to getting a new job or gaining access to the labour market among those who were job seeking.

Refugees also highlighted qualifications, immigration and the asylum process and their lack of information or lack of familiarity with the UK system as barriers. Knowledge and information for new migrants is often acquired through kinship, social and community networks rather than through formal statutory routes. The importance of social networks in

Table 10.3 Main Barrier and All Barriers to the Labour Market

	Main Barrier	All Barriers
Percentages		
English language/literacy	30	48
Lack of UK work experience	19	42
No qualifications	7	25
Awaiting decision on status/immigration status	6	10
Employer discrimination	5	21
Qualifications not recognised	5	12
Unfamiliarity with the UK system	5	24
Lack of information	5	17
Health problems	4	8
Age	3	5
Lack of childcare	3	9
Lack of work in area	2	7
Don't know	2	3
Not ready to work	1	4
No demand for skills	1	2
Lack of confidence	1	7
Lack of time	1	1
None	1	1

Base number: 149

refugee settlement has been found with different cohorts of refugees. Gold (1992), for example, working in the North American context notes that collectivism, that is, ethnic solidarity, provides crucial social, economic and informational resources. In the UK context, before the introduction of compulsory dispersal in 2000 and the episodic dispersal of programme refugees, including Vietnamese, East African Asians and, more recently, Bosnians and Kosovans under temporary protection, refugees have congregated in areas where communities already exist and are able to draw on these social and community networks for support and information in the early stages of migration but also for routes to employment. These networks could be limiting, and research argues that a precursor to integration and social cohesion are social connections and networks within and between communities (Ager and Strang, 2004). When assessing the impact of barriers including accent and discrimination, McKay et al. note that 'such barriers leave refugees with limited options other than to rely on more favourable word-of-mouth methods of recruitment within their linguistic communities' (2006: 127). An area to therefore be explored is the impact of policy, most notably the dispersal of asylum seekers around the country and placements by local authorities outside of the local authority area on social networks and routes to employment.

METHODS OF JOB SEEKING AND SUCCESSFUL ROUTES TO EMPLOYMENT

Job seeking and successful entry routes into employment can be culturally specific. In order to explore job seeking and successful routes to employment, refugees who were looking for work were asked what they had done to find a job, while those who were working were asked how they had found their current job.

Those who were not working and looking for work at the time of the survey were not necessarily looking for work that reflected their premigration employment or their qualifications. In fact, nearly half (47 per cent) said that they were looking for shop/cashier work, factory or dressmaking work or simply 'anything'. Refugees' looking for work that does not reflect their skills levels has been noted in other research (Phillimore and Goodson, 2006). Among those who were unemployed and looking for work there was a real lack of knowledge about statutory provision. Around half (49 per cent) had not heard of any of the schemes run by Jobcentre Plus, and even fewer had participated in any of the schemes, though, as Figure 10.2 shows, 59 per cent had in fact visited the Jobcentre.

Figure 10.2 shows that two-thirds of those who were unemployed and looking for work had asked their friends about jobs. An examination of successful routes to employment shows such a strategy has been more successful than other job search strategies. Among those that were working at the time of the survey, around a third (32 per cent) had found their job through a friend. An additional 7 per cent had found their job through a personal contact already employed at their place of work, and 6 per cent had found their job through a relative. Social and kinship networks were clearly crucial in terms of access to employment. However, the dependence on networks for job search and access to the labour market can perpetuate the unskilled and secondary-sector employment occupied by refugees. Certainly in this study twelve out of sixteen of those working in shops or as cashiers had found their job through networks, as had fifteen out of twenty who worked

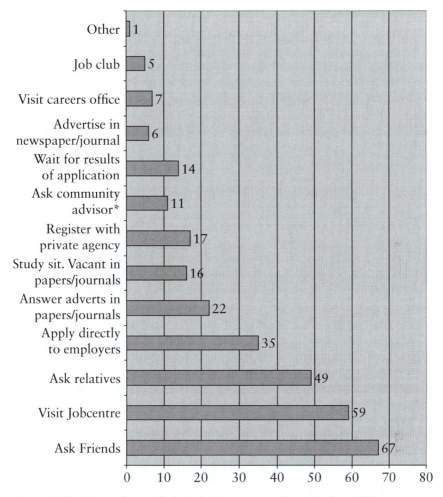

Figure 10.2 Action taken to find a job. Percentages. (Base number: 105).

in catering and seven out of seventeen who were working as interpreters or translators. While informal networks can be important as a stepping-stone, the risk is that employment within coethnic businesses is unlikely to facilitate social contacts outside of the immediate community and therefore will not offer the opportunity for wider social networks which the government argues is key for social cohesion (Zetter et al., 2006). The Commission on Integration and Cohesion noted that,

> An integrated and cohesive community is one where . . . There are strong and positive relationships between people from different backgrounds in the workplace, in schools and other institutions within neighborhoods (2007a: 10).

The importance of employment in the pursuit of social cohesion has been noted (Hudson et al., 2007), as has the interaction between alienation and a lack of opportunities (Commission on Integration and Cohesion, 2007b). Morrice (2007) argues that it can be difficult for refugees to establish the range of social networks associated with economic and social activities including work. The potential for alienation from the wider society through a lack of employment and among those who are working the social limitations of their jobs is certainly apparent. When friendship circles are explored, the picture of refugees located closely within their own community reinforces the difficulties that they might experience accessing these wider networks.

FRIENDS, SOCIAL NETWORKS
AND PATHWAYS TO EMPLOYMENT

The previous section showed the importance of friends and other social networks in both facilitating employment and as a means of looking for work. Before the introduction of dispersal in 2000, refugees tended to live in areas where they either had preexisting kinship and/or social networks or where there was already an established community, so there were variations in the geography of settlement depending on the migration patterns of the different groups and the extent to which there were already existing and established social and community networks. Dispersal during the asylum process has meant that asylum seekers who refused to go to the allocated areas, or absconded, forfeited their right to accommodation and instead rely on family or friends for support—something that is easier for single men than women with children who can not risk losing support (Sales, 2002). In the DWP study 7 per cent of respondents had been dispersed to their current areas of residence and 16 per cent had been placed by another local authority, so in total just under a quarter did not decide on their place of residence. This section will explore whether there were any differences in terms of employment profiles and pathways to employment among those who had a choice where they lived and those who did not, either through dispersal or placement in a locality by another local authority.

In order to explore the use of social networks in employment, the research set out to establish more about why people lived in the area that they did and friendship circles. Research consistently shows that refugees can have limited or no say in their country of asylum and that agents and smugglers are influential in determining refugees' final destination (Robinson and Segrott, 2002; Gilbert and Koser, 2006). This means that some refugees have limited networks on arrival to the UK and certainly initially in their area of residence. The migration patterns of different groups interacted with reasons for living in an area. For example, people from Somalia and Turkey (including Kurdish people), with their long history of migration to the UK, were most likely to be living in an area because of their kinship, social and community networks. In contrast, refugees from Kosova, characterised by their comparatively recent migration and lack of established communities, were least likely to have exercised choice over their area of residence, as Table 10.4 shows. The ways in which these factors played out in terms of employment will be examined.

In terms of friendships and community activities differences emerged between refugees from different countries of origin, though nearly everyone had made new friends since coming to the UK (96 per cent). More than half said that their new friends were from their own community (59 per cent); 35 per cent said that their friends were from a mixture of different groups; and the rest (6 per cent) said their friends were mainly refugees from other groups, mainly people from other minority ethnic groups or mainly 'white' British people. An interesting finding was the lack of variation in terms of the new friendships among those who lived in an area due to family, social or community networks and those who were dispersed or placed in an area. The research was carried out only a few years after the introduction of dispersal and took place in urban centres, which could have affected access to networks from the same country of origin. There was, however, a difference in the profile of new friends among refugees who lived in an area for employment or educational purposes. This group were much more likely to have a mixture of new friends (60 per cent) than those who lived in an area because

Table 10.4 Main Reason for Living in Area by Country of Origin

	Frequencies					
	Somalia	*Turkey*	*Iraq*	*Sri Lanka*	*Kosova*	%
Family	37	36	12	20	18	31
Friends	18	13	8	14	12	16
Got placed by local authority	14	2	2	3	41	16
Community	2	14	19	4	2	10
Employment/education	2	3	15	18	2	10
Dispersal	2	5	13	5	3	7
Like the area	1	0	5	9	0	4
Cheap housing	1	1	0	4	0	2
*Other	3	6	5	3	2	4
Total	80	80	80	80	80	400

*Other includes places of worship, services for refugees, didn't know anywhere else.

of their networks and those who had no choice (33 per cent and 26 per cent, respectively). This difference feeds into the social-cohesion arguments that interaction at work and in other institutions, including educational establishments, can offer routes to integration, though such a simplistic analysis fails to explore the interaction of other important variables, particularly language, which is a key facilitator of mixed networks. English-language proficiency was very important in determining the profile of new friends. While 43 per cent of those who spoke English fluently or fairly well had a mixture of new friends, among those who spoke English a little or not at all the proportion was 13 per cent. In fact, language rather than the reasons for living in an area was the factor that most determined new friendships.

Living in an area due to networks and community did increase the likelihood of paid employment—26 per cent of those living in an area because they had social and community networks were working, compared with 16 per cent of those who had no choice, and it also affected the ways in which people had found their current job. Table 10.5 shows the ways in which people had found their current or most recent job in the UK. Though the numbers are small among those who had no choice about their locality, the data still suggest the importance of friends in terms of pathways to employment, though friends and personal networks were more apparent among those who had chosen to live in an area because of these networks.

Table 10.5 Ways of Finding Current or Most Recent Job by Reasons for Living in Area (Base Number 114)

Means of Finding Current or Most Recent Job	Living in Locality Because of Social & Community Networks		No Choice in Locality	
	Number	%	Numbers	%
Advert	17	20	—	—
Jobcentre Plus	6	7	4	13
Agency	2	2	6	20
Direct application	9	11	5	17
Own business	2	2	1	3
Friends/knowing people already working there	39	46	11	37
Relatives	5	6	—	—
Community group	4	5	3	10
Total	84	100	30	100

One of the key findings of the study in terms of policy analysis was the lack of take-up of statutory job search provision. Table 10.5 shows that a larger proportion of those who had no choice about where they lived had used either an employment agency or Jobcentre Plus to find work. Since the study was carried out, the importance of early contact with Jobcentre Plus has been highlighted, and the introduction of a 'voluntary marker' in 2004 has been introduced to try to better monitor contact with and outcomes for refugee-service users (see DWP, 2005; Employability Forum, 2006).

The immersion in networks for routes to employment can perpetuate the tendency for refugees to be employed in secondary-sector jobs, and an examination of jobs-search strategies suggests that this pattern will continue to replicate itself. Even if refugees use Jobcentre Pus more regularly, the jobs available often lack opportunities for progression and will not further career pathways for refugees.

Those refugees who were looking for work were asked what they had done in order to find a job. Not surprisingly, those who had new friends mainly from their own community were more likely to ask friends and relatives for information about jobs than those who had a mixture of friends. However, Figure 10.3 shows that refugees who had no choice about where they lived were more likely than those who were living in a locality due to social and community networks to use both statutory and informal modes of job search.

The data suggest that dispersal and placement did not remove the possibility of informal networks as a method of job seeking for refugees, but it also resulted in a greater use of statutory provision, neither of which will facilitate jobs that are commensurate with skills and qualifications among highly skilled, skilled and professional people.

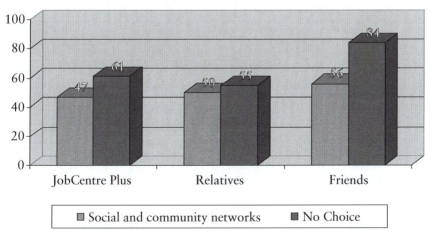

Figure 10.3 Methods of job seeking by reasons for living in area. (Base number: 114).

CONCLUSION

This chapter has shown that refugees arrive in the UK with diverse skills sets, though some are highly skilled on arrival or have achieved new qualifications in the UK. Almost uniformly, refugees are unable to use their skills and experiences and are working in secondary-sector jobs, for low pay, with poor terms and conditions of employment and with little or no opportunity for progression. The reasons for this are complex and include the limited networks that refugees have and their heavy reliance on these networks for job seeking, which perpetuates and continues to root many in secondary-sector jobs. Even those who are fluent in English, though more likely to work, are for the most part employed in secondary-sector jobs. Other research has highlighted employer concerns about checking and understanding documentation, attitudes to refugees, discrimination and accent as impeding refugees' opportunities (Hurstfield et al., 2004; Archer et al., 2005; McKay et al., 2006).

Refugees remain firmly fixed within their own communities, regardless of their reasons for living in an area. Most new friendships are with others from the same country of origin, though English-language fluency, which is linked to employment and participation in education, does help to extend people's social networks and facilitates a more diverse set of new friendships. Dispersal and placement by local authorities have not impeded new friendships with people from the same country of origin or the use for these networks for job search. Jobcentre Plus is trying to facilitate better take-up of its services among refugees, though their work tends to focus on nonprofessional employment. It is widely recognised that statutory provision cannot assist professionals or support long-term retraining. Though some schemes are in place to help professionals from refugee backgrounds, this is an area that remains a challenge (Employability Forum, 2006).

The research reported on in this chapter shows that refugees experience basic barriers to economic and social integration. In terms of economic integration, some barriers relate to personal capacity, though not exclusively, as some refugees with English fluency and high skills are not achieving their potential and not gaining employment appropriate to skills and ability—a prerequisite for integration (Home Office, 2005). Information provision, local areas and community structures, government policy as well as employer discrimination all contribute to the complexity of refugees' experiences in the labour market and their disproportionate levels of unemployment and underemployment. To facilitate refugee integration by ameliorating barriers to the labour market and appropriate levels of employment, a complex set of measures and new practices need to considered that spans the range of disadvantages and discriminations experiences by refugees in the UK.

NOTES

1. The term 'refugee' will be used to describe all forced migrants (refugees, asylum seekers and those with humanitarian status, unless a distinction is specified).
2. See A. Bloch, (2004a) for a detailed report on the methodology.
3. Some of the data used in the chapter have been used in a different form in Bloch (2002) and Bloch (2004b).

11 Immigration and Labour Market Integration

Anne E. Green

The role of immigrants in the labour market is receiving increasing public-policy attention in the UK. The UK government has increasingly emphasised the development of a 'managed migration policy' designed to meet the needs of the UK economy (Somerville, 2007) with migrants playing an important role in filling labour shortages and addressing skill deficiencies. There were a number of managed migration routes for migrants from beyond the EU. These include the Work Permit system (for non-European Economic Area migrants filling specific vacancies), the Highly Skilled Migrant Programme and a number of special schemes focusing on specific sectors at the lower end of the labour market where posts have proved difficult to fill (including Sector Based Schemes and the Seasonal Agricultural Workers Scheme). In 2006 a points-based system (PBS) for managing labour migration was announced (Home Office, 2006), to be rolled out in 2008. The PBS is designed to maximise the economic contribution of migration, under the guidance of a Migration Advisory Committee, which will advise ministers on where migration might sensibly fill gaps in the labour market.

However, not all migration flows to the UK are 'managed'. Alongside immigration through the channels outlined earlier, the asylum route has proved an important avenue for immigration to the UK, albeit one that has varied in importance over time. There was an upsurge in entries of people to the UK via the asylum route from the late 1990s through to 2002, prompting new policy initiatives in relation to the settlement and entitlements of asylum seekers, and measures to tackle asylum abuse. However, since 2003 the number of asylum seekers to the UK has fallen.

Another key source of non-'managed' migration involves individuals from other European Union (EU) member states. Since 2004 particular attention has focused on migrant flows from the new Member States of Central and Eastern Europe (notably Poland). This new wave of economic migrants has been significant in terms of their geography—they display a more spatially dispersed distribution than many former waves of immigrants (Commission for Rural Communities, 2007); their occupational profile—they are concentrated to a greater extent in less skilled occupations than immigrants in aggregate; and their future intentions—they are characterised by a greater degree of uncertainty about their length of stay in the UK, coupled with a greater ease of travel back to their country of origin, than some other migrant groups.

This discussion of 'managed' and 'nonmanaged' flows, and of some of the different types of 'nonmanaged' flows which have risen to prominence in recent years, underlines the point that the profile of immigrants to the UK is subject to changes in volume and composition over time. Some of the barriers to labour market integration and associated support needs are common to immigrants generally, while others are more specific to particular 'groups', albeit the heterogeneity of individuals within groups means that it is often difficult to generalise group experiences in a meaningful way. Likewise, some initiatives for promoting labour market integration are of universal applicability across 'groups', while others may have a narrower and more specific focus.

This chapter addresses the issue of how to maximise the economic contribution of refugees and economic migrants by fostering labour market integration. In particular, it draws on research focusing on refugees in London and on economic migrants in the East Midlands arid West Midlands regions of England.

The chapter is organised into three substantive sections. The first section sets the migration, labour market and governance context for labour market integration initiatives. Here the key points are that relatively favourable economic performance in the UK (particularly vis-à-vis EU competitors) in recent years, along with the flexibility and openness of the UK labour market, has meant that the UK has been an attractive destination for migrants. However, while the UK government sets a national framework for migration policy (which, as outlined above, is the subject of ongoing change), many issues of labour market integration are dealt with by a patchwork of institutions and agencies at regional and local levels.

The second section highlights key barriers to labour market integration, and so to maximising the economic contribution of migrants at local, regional and national scale. Two groups of key harriers are highlighted: first, harriers to labour market participation; and secondly, barriers to labour market advancement. Some barriers such as language, information and knowledge on how the labour market operates, and qualification and skills recognition, are common to both participation and advancement; whereas others relate solely to either participation of advancement. Both supply-side and demand-side considerations, and an appreciation of the local institutional context, is important in understanding how barriers operate in practice.

The third section of the chapter is concerned with initiatives to facilitate labour market integration in different local contexts. Selected examples of the scope, nature, strengths, weaknesses of local initiatives to help integrate refugees and economic migrants into the labour market are presented. Many of these focus on labour supply. Increasing numbers are concerned with the activities of labour market intermediaries and the local institutional context, while relatively few address labour demand. Some of the challenges in assessing 'what works' from a multiplicity of different policy interventions at local level are outlined.

The chapter concludes with a discussion of some key themes and future directions emerging from the preceding discussion.

CONTEXT

Migration Context

Migration to the UK is not a new phenomenon, although it has been running at historically high levels in recent years. The size, character, nature, origins and destinations of migration flows have varied over time, as outlined earlier, and the specific characteristics of migrants entering the UK via different migration routes have implications for labour market integration. At times of particular 'crises' some specific policy responses have been framed in a 'reactive' fashion, so leading to an increasingly complex picture over time. In turn, this has led to calls to 'rationalise' and 'simplify' the legislative context for the migration system so as to reduce complexity, enhance transparency and ensure that it serves key government objectives. The announcement of a five-year strategy on asylum and immigration in February 2005 (Home Office, 2005a) exemplifies this by proposing that refugees should no longer be granted immediate settlement in the UK, but should instead be granted five years' temporary leave. This can be viewed in this broader context of a need for stocktaking, simplification and a reaffirmation of primary policy objectives.

As outlined in the Introduction, the UK received an increase in asylum seekers from the late 1990s. Prior to this time, the UK had received relatively few asylum claims, and a high proportion of those seeking asylum were awarded refugee status. Factors prompting this change in the 1990s included the collapse of communism in Central and Eastern Europe, the emergence of failed states in other parts of the world, reductions in the price of international air travel, the development of highly organised trafficking networks linked to organised crime, and also a greater awareness in developing countries of the relative prosperity in countries such as the UK. Key features in the relative attractiveness of the UK as a destination were a relatively strong economy offering a ready supply of work, the global reach of the English language and the perception of the UK as a tolerant society.

As the numbers of asylum seekers rose, especially in the period from 2000 to 2002, many of those seeking asylum in the UK were held not to have protection needs. The government responded by taking a number of measures to deter unfounded asylum applications. One reform from this period of crucial importance, from a labour market perspective, was the withdrawal of the employment concession, whereby asylum seekers who had not had an initial decision within six months had a right to work, in July 2002. This was done because it was considered that access to the labour market was acting as a 'pull' factor, encouraging economic migrants to claim asylum in the UK. However, it is government policy to support labour market integration of successful asylum applicants.

Migrants entering via other routes do not face such restrictions. Free-movement rights mean that citizens of the EU15 (i.e. the first fifteen European Union Member States) do not need permission to work in the UK. With EU expansion to Central and Eastern Europe in May 2004, the UK put in place transitional measures to regulate access to the labour market by nationals of the 'A8' countries via the Worker Registration Scheme. This expansion of the EU in 2004, and the fact that the UK was one of only three member states (alongside Sweden and Ireland) that chose not to impose restrictions on so-called 'Accession 8' (A8) migrants, has led to much greater than expected numbers of migrants from new member states in Central and Eastern Europe coming to the UK since 2004. Indeed, the scale of recent immigration is such that Poles have been identified as the largest ever single national group of entrants that the British Isles has ever experienced (Salt and Millar, 2006). With the accession of Romania and Bulgaria to the EU in January 2007, additional restrictions were placed on migrants from these 'Accession 2' (A2) states.

Some migrants enter the UK illegally, and others enter legally but work illegally. There is an unknown number of failed asylum seekers living in the UK (estimated by the Home Office [2005b] to be 430,000 and by Migration-watch [2005] to be 670,000) who are not able to work legally, but may provide a source of undocumented workers. No particular attempts are made in this chapter to ascertain the size of, or gather information specifically on, undocumented migrants. Likewise, trafficking for purposes of labour exploitation is not specifically covered here, although it is noted that recent research has shown that typically victims are found in industries requiring large numbers of low-paid, flexible, seasonal workers (Dowling et al., 2007).

Labour Market Context

UK labour market policy since 1997 has had a strong supply-side bias. Labour market participation has been regarded as *the* key route out of social exclusion. There have been many initiatives to enhance labour market participation and employability. The New Deals (focusing on the longer-term unemployed and excluded groups) form the centrepiece of welfare-to-work initiatives. In order to 'make work pay', a national minimum wage has been introduced, as have a series of 'in-work benefits'.

More specifically, a range of levels of support to enter the labour market is available to refugees and economic migrants, while asylum seekers lie outside the system of labour market support following the withdrawal of any right to work. There is a lack of comprehensive data on the relative position of immigrants, and specifically refugees in the labour market, and on the relative success of different labour market initiatives. This reflects a lack of evaluation evidence from many small-scale (and larger-scale) projects focusing on refugees and other immigrant groups. However, from April 2006 refugees have been recognised as a priority group by Jobcentre Plus, and so the evidence base on participation and outcomes from some mainstream programmes is set to improve.

In the case of refugees not currently in employment, the strategy for labour market integration rests on encouraging more refugees to use Jobcentre Plus and by building stronger partnerships involving the voluntary and community sectors at local level—including Ethnic Minority Outreach work involving intermediaries, and delivering services that meet the needs of refugee customers. Initiatives of specific relevance to refugees include SUNRISE (Strategic Update of National Refugee Integration Services) to provide a dedicated caseworker to help and support individuals and an expansion of Time Together mentoring (National Refugee Integration Forum Employment and Training Subgroup, 2006). These developments represent a cross-government strategy, requiring coordination across government and a wide range of support agencies, designed to ensure that refugees are able to make 'a full and positive contribution to society'. It is recognised that integration is a complex process and that its achievement can be measured in many different ways. The strategic policy framework focuses primarily on integration through employment, and so with this in mind, integration can begin in the fullest sense only when an asylum seeker becomes a refugee.

For economic migrants such specific integration programmes do not exist. For those entering the UK via the work-permit route, an employment position is established at the outset. Some economic migrants with free-movement rights set up employment before leaving their home country; others use agencies, social networks and other recruitment channels to find work once in the UK. Although benefits for some groups of migrants are initially restricted, most come under the auspices of Jobcentre Plus services to aid labour market integration.

GOVERNANCE

Although the legislative framework for migration is set at national level, the governance framework at regional and local levels provides an important context for the implementation of migration and labour market integration policy in the UK. Overall, there is a patchwork of national, regional and local actors responsible for different aspects of migrant integration—including central government departments and agencies, regional agencies, local government and nongovernmental organisations. Alongside these, the community and voluntary sector (including church-based organisations) also play an important role. The number of stakeholders potentially involved in providing support, and the number of instruments used, makes for a complex picture and poses a significant governance challenge. Problems are particularly acute in relation to funding of activities for refugee/migrant labour market integration. More generically, the factors which can prevent effective support being given to refugees and economic migrants may also undermine wider policies to promote economic development and social cohesion (as outlined following). These can include poor communication and coordination between stakeholders, a lack of integration between supply and demand, a poor prioritisation of resources, an absence of long-term local strategies, an emphasis on short-term impacts rather than long-term change, and a grants-based culture of provision which does not encourage either mainstreaming or sustainability.

Many of these problems are compounded by the fact that the number of refugees and migrant workers at local level cannot be estimated with any confidence. Attempts have been made at local level to estimate numbers of migrants and refugees using a range of secondary data sources, administrative records and information from service providers, but any such estimates need to be treated with caution and may date very quickly. In addition to data on numbers of economic migrants and refugees, information on the characteristics of migrants and refugees, and so the labour market integration challenges they pose, is needed (Audit Commission, 2007). Historically, migrants have tended to move to urban areas—especially to certain 'gateway' cities. However, as outlined at the beginning of this chapter, some of the new migrant groups (notably A8 migrants) display a more 'rural' distribution, such that local areas with less experience of dealing with migrants and refugees, and with a less developed integration 'competence' and infrastructure, are facing challenges of the labour market integration of immigrants for the first time.

In summary, in the context of an unprecedented rise in the number of refugees and asylum seekers since the early/mid 1990s, with a peak in the early years of the twenty-first century, and a subsequent influx of economic migrants from Central and Eastern Europe, the policy environment for regulating migration and enhancing integration to the labour market has been volatile. Key parameters of government policy in relation to refugees and asylum seekers and economic migrants over the last decade, of specific relevance to the concerns of this chapter, include the ongoing development of a 'managed migration' programme for economic migrants in relation to labour market needs and skill shortages (as outlined earlier); deterrence of unfounded asylum claims; accelerated and strict status determination procedures and development of policies for integration of refugees and asylum seekers with leave to remain (see Zetter and Pearl, 2005).

BARRIERS TO LABOUR MARKET INTEGRATION

All job seekers—whether or not they are refugees or economic migrants—face barriers to labour market integration. However, specific aspects of more general barriers may pose greater problems for immigrants than for other population subgroups. This section is concerned with identifying general and more specific barriers to labour market integration of migrants. Barriers to both labour market participation and advancement are considered. At the outset, however, it is salient to note two general points. The first is that labour market integration represents only one aspect of broader integration, albeit a very important one economically, socially and psychologically for most migrants. The second is that challenges that refugees and economic migrants face, in entering and moving within the UK labour market, are many and varied, reflecting their heterogeneity.

Barriers to Labour Market Participation

As outlined at the beginning of this chapter, *legal barriers* may place obstacles for some job seekers in terms of their availability for work (as in the case of asylum seekers who have not yet been granted refugee status). In instances where professions/other positions are subject to regulation in the UK, it may be the case that immigrants are not permitted to work in such jobs until they have acquired requisite qualifications and experiences in the UK. Immigrants may become concentrated in the informal economy where they have entered the country illegally and/or where the barriers to entry to the formal economy are difficult to overcome.

Barriers associated with *availability for work* are likely to be a more pressing issue for refugees than for economic migrants, whose prime motivation to come to the UK is to seek employment. Care responsibilities (for children and adults) and a lack of access to appropriate and affordable care provision may place limitations on the availability for work of some potential job seekers. Since refugees (and economic migrants) are less likely to have relatives in the local area than the more established population, they may find it particularly difficult to delegate caring responsibilities. Social, psychological and motivational problems may also place limitations on an individual's availability for work. Some refugees may face additional emotional problems in overcoming disorientation and trauma before they are in a position to look for work. For some a loss of identity and sense of self-esteem, coupled with low motivation—due to long periods out of work and training, loss of employment-related networks, feeling discriminated against, and feeling unwanted and outcast—may contribute to further lack of confidence. Conversely, some groups of economic migrants are highly prized because of their motivation and strong work ethic (Green et al., 2007). This underlines the diversity between different groups of immigrants.

Even if immigrants have a legal right to work in the UK and are available for work, deficiencies in *information and knowledge of how the UK labour market works* may be a barrier to labour market participation for some immigrants. Those immigrants (and other job seekers) who have deficiencies in *generic skills and competencies*—most notably communication, but also reliability, knowledge of how to behave in the workplace, ability to follow instructions and to work as part of team alongside others—that are required in almost all jobs are likely to face particular challenges in finding work.

In some instances a lack of information and knowledge and deficiencies in generic skills may be compounded by poor *English-language skills*. The literature on migration highlights the importance of English-language skills for labour market and wider social integration. Moreover, it is important for individuals' self-sufficiency and for navigating one's way around the labour market. The ability to speak English is not essential for all jobs, but it is likely to be a particular issue in customer-facing roles or when technical jargon is used. Research on migrant workers in the West Midlands highlights that employers identify the existence of a language barrier as the main disadvantage of employing migrant workers (Green et al., 2007).

Reference has been made to employers already in relation to English-language skills. *Employer attitudes* play a role in shaping whether or not immigrants are able to participate in the labour market and the nature of such participation. Any job seeker may face prejudice on the basis of social background, place of residence or other stereotypical characteristics. Immigrants, particularly from visible ethnic minority groups, may face racism. Moreover, the concentration of refugees in poorer areas, which may become stigmatised, can add to problems faced in accessing employment. Confusion amongst employers about who is and who is not permitted to work, and about the complexities of clarifying permission-to-work documentation, may also be a barrier to work for refugees, especially vis-à-vis some other groups of economic migrants. Negative media reporting about refugees and other migrants also impacts on perceptions of some employers (IES, 2004), such that they may be unwilling to recruit such job seekers. On the other hand, as noted previously, some employers may have a preference for some groups of migrant workers to undertake some types of jobs because of the generic skills they offer. Hence, employer attitudes may be a barrier to labour market participation of some immigrants, but not for others.

Barriers to Labour Market Advancement

Barriers to full labour market integration are faced not only by those outside the labour market, but also by those who are in employment. Many of the issues identified earlier in relation to labour market participation also pose barriers to labour market advancement. *Language* provides a good example. Shortcomings in English language emerges repeatedly in research studies (see Bloch, 2002) as a barrier to participation to mobility within the labour market; for many immigrants, better English is a prerequisite for improving their employment position (Audit Commission, 2007). In the context of the restructuring of employment in the UK with an increasing share of employment in the service sector and greater emphasis on customer-facing roles, there are few jobs in the UK that do not require at least a limited grasp of

English. It is clear that English-language skills enhance individual employ-ability and earnings (NIACE, 2006). Many migrants recognise this, and some explicitly seek work roles where they can practise their English (for example, in sales and customer-service roles). A study of migrant workers in the West Midlands (Green et al., 2007) indicated that migrants self-assessed their skills in understanding and speaking English as better than those in writing English. These poorer writing skills may limit the options of those who wish to study for UK qualifications.

In some instances understanding the *channels and mechanisms via which immigrants are recruited* into employment sheds light on the nature of sub-sequent barriers to labour market advancement. Those who entered the UK on work permits were least likely to face barriers to labour market advancement, because they have been recruited to undertake a specific job. Other migrants are more likely to face barriers. Agencies (either in the UK or in origin countries) are a widely used recruitment mechanism amongst economic migrants. However, research in the West Midlands suggests that once established in the UK, some migrant workers with higher level skills that are not utilised in the operative and elementary occupations that typify much agency employment are keen to use social networks to advance in the labour market (Green et al., 2007). For economic migrants the initial attraction of agency work is likely to be ease of labour market entry at a time when they need income in order to live in the UK. These same typical operative and elementary jobs often also provide the easiest 'ports of entry' to the labour market for refugees. However, the way in which some of these jobs are organised and their *time-space dynamics*—particularly those with long hours for low pay, changing shift patterns, and sometimes changes in workplace location at short notice—mean that it can be difficult to advance out of such jobs. This is especially so when working arrangements inter-fere with participation in training that may be necessary for labour market advancement. Hence, some migrant workers/refugees may work long hours in poorly paid jobs, such that they face financial hardship and/or a lack of time, opportunity, energy or inclination to engage in further training with a view to moving upwards within the labour market. In such circumstances, individuals can easily end up devoting their energy to living from one day to the next, rather than strategising for a better medium-/long-term future.

Refugees and economic migrants engaged in low-grade jobs which do not utilise their qualifications and experience may suffer poor motivation, as well as some redundancy in skills which are no longer practised. In many instances barriers associated with *qualification and skills recognition* may be of greater significance in terms of labour market advancement than labour market partipation per se, because individuals can often participate in the labour market, but not in their preferred occupation or at a level commen-surate with their qualifications, skills and experience. In order to find work that best suits their skills and aspirations, job seekers require *skills in navi-gating the labour market*. Here, lack of knowledge of the UK system and of how local labour markets operate, and a lack of familiarity with local social networks, may place refugees and migrants at a disadvantage vis-à-vis the indigenous population. In some cases a lack of UK work experience may be an important issue, since employers often lack knowledge about the value and relevance of qualifications and experience gained outside the UK. From an individual refugee/economic migrant perspective such a situation means that individual aspirations are not being fulfilled, while from a broader economy perspective it means that the economic contribution of immigrants is not being maximised.

This raises the question of whether individual migrant workers/refugees want to advance in the labour market and also whether employers want them to do so. Some migrants intend to stay for a relatively short period only, whereas at the opposite end of the spectrum some migrant workers and many refugees intend to settle permanently. On the basis of evidence from migrant worker and employer surveys in the West Midlands (Green et al., 2007), it appears that many employers want classic 'economic migrants' (i.e. workers who are attracted by higher pay in the UK than in their origin country, but who are content to stay in low-skilled roles which employers would otherwise find hard to fill), while many migrant workers 'aspire' to improve their skills/improve their language abilities/gain other relevant qualifications s so that they can advance in the labour market—either in the UK, their country of origin or elsewhere.

INITIATIVES TO FACILITATE LABOUR MARKET INTEGRATION

Context: Introduction to Types of Generic Support

In the light of the generic issues and barriers to labour market integration highlighted earlier, refugees/economic migrants are likely to have a number of support needs, if their labour market integration is to be facilitated. These range from activities requiring a relatively limited input of resources to much more resource-intensive initiatives.

On the supply side, where the majority of initiatives to facilitate labour market integration are focused, and starting at the less resource-intensive end of the spectrum, *help with job search and CV preparation* can assist those who have skills and competencies to access employment but who lack knowledge about the practicalities of job search and who are unfamiliar with the language of job hunting in the UK and lack knowledge of how vacancies occur and how they are publicised. More *general familiarisation with the range of employment opportunities available*, equal opportunities and employment legislation in relation to job hunting and in the workplace, the level of competition for jobs, workplace practices/procedures/jargon, how to describe 'home' qualifications, transferability of skills, and how to access training in the UK are also likely to be valuable. Some individuals may require *help with social networking* to gain access to appropriate job openings. *Mentoring* may be valuable here. Moreover, successfully established immigrants may be valuable role models for new arrivals and prove positive in boosting confidence, providing advice and sharing experiences.

For some refugees/economic migrants, *specific training* may be necessary to enable labour market participation and advancement. In some instances what may be required is short-term training for specific vacancies. Support needs in terms of *language training* will vary from individual to individual, taking account of existing English-language ability and general educational levels. Some require very specific work-focused and intensive provision, while others require ESOL (English for Speakers of Other Languages) at preentry level alongside basic skills support. Existing ESOL provision has been identified as being 'beset by enormous problems' (NIACE, 2006) and is of variable nature and quality (KPMG, 2005).

Work experience placements/programmes may be valuable in some instances for some individuals—particularly as part of a broader support programme. However, experience of local projects suggests that persuading and arranging for employers to take refugees on work placement for work experience is time consuming and difficult. *Qualification recognition* is another relatively resource-intensive activity, but one which is essential for migrants to make optimal use of their skills. Moreover, from an economic perspective, such recognition can show the extent to which migrants have qualifications (and skills) relevant to shortage sectors. To date, the facility for benchmarking and UK equivalence of qualifications is better for academic qualifications (via the National Academic Recognition Information Centre for the UK [UK NARIC]) than for vocational qualifications. Moreover, confining discussion to 'qualifications' tends to discount the 'skills' (not necessarily captured by 'qualifications') that refugees/economic migrants have. Qualifications are easier to compare and to map than are skills, yet the importance placed by many employers on 'soft' skills and relevant work experience alongside qualifications remains an important issue facing migrants to the UK, suggesting that except in sectors/occupations facing very severe shortages, qualifications alone may not be enough to enable an individual to access employment at a level commensurate with his/her skills.

Assessing 'What Works': Examples of Local Initiatives

Evaluation Challenges

The support needs identified earlier illustrate some of the spheres of possible activity on the supply side of the labour market of relevance to the labour market integration of refugees/economic migrants. However, assessing 'what works' in policy terms is difficult for a number of reasons relating to both the nature of provision and the nature of refugees/migrants (Hasluck and Green, 2007). Considering the *nature of provision*, first it is clear that interventions can take many different forms, such as advice, guidance, work placement, language training, vocational training, basic skills training and so on, but in any one such category provision is likely to be heterogeneous in terms of content, timing and context of delivery, such that what might work in some contexts may not work in others. Secondly, provision is often packaged into a project or programme containing several 'ingredients' and/or individuals may be recipients of a number of different policy interventions, such that it is difficult to isolate the impact of one ingredient working without several other ingredients being present to complement it. Moreover, the 'right' combination of ingredients is likely to vary between individuals.

Given the considerable diversity between, and intragroup heterogeneity amongst, different groups of immigrants, it is likely that what is appropriate and 'what works' for one individual may not work for another. Moreover, attempts to assess 'what works' are often thwarted by *limitations in the evidence base*. Overall, a tendency for a reliance on short-term funding and an emphasis on 'innovation' in the introduction of new initiatives (often to the detriment of building clearly upon the experience and 'good-practice' elements of existing programmes) has contributed to a general lack of robust evaluation evidence.

Local Initiatives

Access to education and training (albeit that individuals vary in their education and training requirements) is fundamental to the labour market integration. It is generally acknowledged that the UK has a relatively flexible and open training system, offering opportunities for lifelong learning after the end of compulsory education. In theory, this should act to the advantage of immigrants. Yet flexibility and openness can mean that it is difficult to gain a full understanding of how the system works, and to select appropriate courses. In order to fill this information gap for refugees, the Refugee Education Training and Advisory Service (RETAS) has produced a range of guides and services to refugees in the UK—including general overviews of the education system and sources for further information, and a range of advice and training services to help refugees access education, training and employment (including self-employment). On a smaller scale, and at a more local level, many third-sector organisations aim to facilitate access to training through specific training projects involving vocational and language skills (often at entry level) for those who may well be unwilling or unable to access mainstream provision—at least in the first instance.

With the increase in immigration to the UK, demand for *English-language support* has outstripped supply, despite increases in the expenditure on ESOL and in the volume of provision. In major cities, and especially in London and other established centres of migrant concentration, there is a wide range of ESOL provision—delivered by colleges and by third-party organisations in community settings, but questions remain regarding the extent to which provision matches the needs of learners and the demands of the economy—in terms of level (introductory/intermediate/special-ist), content (specifically in relation to fulfilling workplace needs—which will vary by sector and occupation) and availability (geographically and temporally).

Experience suggests that work-related language courses, in combination with adaptation programmes and work placements as part of a specialist package, work well, furnishing individuals with sufficient language ability to function in a specific workplace environment as quickly as possible. In October 2007 a new set of 'ESOL for Work' qualifications were launched. As the title suggests, the new qualifications are shorter and more work-focused than traditional ESOL qualifications. As such, it is hoped that the new qualifications will help employers benefit from improved communica-tion skills and productivity, and that this will encourage employers to con-tribute to the cost of training. The Holbeach campus of the University of Lincoln in the East Midlands exemplifies good practice in such contextual-ised ESOL provision for companies in the food sector. ESOL is provided on a part-time basis at employers' premises, and courses are tailored specifically to employer needs. All courses are contextualised to a particular factory, a needs analysis is conducted, pictures are taken in the factory of signage, products and processing activities, recordings of instructions are made and documentation such as forms, labels and instructions are incorporated into teaching materials (Bowser, 2007).

As outlined earlier, *qualification recognition* (and training) is essential to ensure that refugees and other immigrants are getting the support, infor-mation and advice they need to fully integrate into the UK employment market. As such, access to qualification recognition services can form an essential component of other projects, involving broader labour integration aims than recognition of qualifications per se. The Migrants and Refugees Qualifications project in London—aiming to identify and support transfer-able skills of migrant workers (including refugees) and supporting them to fulfil their potential in the employment market—is one such example (see Green, 2006). Migrants proceed through the project in four sequential stages. The first stage is concerned with 'identification' of eligible clients (often through Information Advice and Guidance Partnerships), onward referral to UK NARIC to complete a Migrant Skills Questionnaire (cover-ing personal details, employment history, qualifications, English language proficiency, etc). In the second 'qualification comparability' stage, NARIC completes qualification comparability documentation and informs the client accordingly. The third stage is 'skills analysis', in which each client receives from NARIC a document that 'maps' the additional training/learning needed to be qualified and competent to UK industry standards. The fourth stage embraces 'local guidance, support and active brokerage', involving one-to-one counselling and guidance, production of a personal development plan, brokerage of additional skills (basic ESOL, vocational, academic, etc), and arranging work experience.

Although refugees and economic migrants often need help to navigate the UK labour market and would benefit from assistance in *acculturation to UK society and the labour market,* such assistance has remained relatively underdeveloped until recently. An example of a project that does attempt to provide an element of acculturation to the UK labour market, alongside other education, training and work experience is the 'Bridge to Work Project', part of the Continuing Education and Training Service (CETS) in Croydon. The project aims to get students with overseas qualifications into work in their chosen profession, establish the equivalence of their qualifications, build confidence, improve interpersonal and language skills, introduce students to UK work culture and explore cultural issues, provide a work placement, fund further training and/or requalification, develop IT skills, provide careers information advice and guidance, and to provide assistance in all aspects of job search. Another example of good practice, this time aimed specifically at migrant workers, is the *Migrant Gateway* portal—http://www.migrantgateway.eu/—developed by the East of England Development Agency and partners. This site is designed to provide information and guidance for migrant workers (those already in the UK and those thinking of coming to the UK), employers, trade unions and other agencies who work with them. The aim has been to bring together information (from official and other sites) about legal information, migrants' rights as employees/workers, rights to housing, immigration status, healthcare, and so on.

Initiatives providing support in *accessing employment* tend to be best developed for refugee professionals. It might be expected that refugees and other immigrants would find it easier to get into unregulated professions (such as engineering) than into regulated professions (such as medicine), but this is not necessarily the case. The advantage of an unregulated profession is that it is not absolutely necessary for a migrant to go through a lengthy and costly requalification process; rather, it is possible to look for work straight away. However, the disadvantage of an unregulated profession is that the path to employment is not necessarily clear, and depending on the migrant's previous experience, can be complex and unpredictable. RETAS provides advice and support for professionals aiming to enter unregulated professions (such as engineering) and regulated ones (such as medicine).

Some refugees and economic migrants facing difficulty in fulfilling their aspirations in the UK labour market have opted for self-employment. Local and regional development agencies are proactive in *encouraging and supporting entrepreneurship*, as a means of helping to sustain and increase wealth and prosperity, while at the same time proactively tackling economic disadvantage. Some immigrants have access to programmes which have a specific focus on providing assistance to ethnic minorities, in addition to more generally available enterprise and financial-assistance initiatives.

The success (or otherwise) of initiatives of the types outlined previously rests not only on the extent to which they achieve their goals, but also on whether they are willing and able to successfully '*signpost*' immigrants to more relevant and/or further support provided by other organisations and to assistance in other spheres relevant to integration—such as health, housing, and so on. This calls for strong *partnership working* and awareness of other activities and initiatives, which takes time to develop and which may be hindered by differences in organisational target structures. In some large cities there is a multiplicity of local projects/initiatives to support refugees/economic migrants. Often local initiatives have grown up organically in a piecemeal manner, leading to an (arguably overly) complex system. The result is that it is difficult for anyone (including those concerned with strategy formulation and delivery, let alone newcomers trying to identify pathways and navigate their way around the system) to gain a clear view of the overall scope of policies and initiatives designed to aid integration into the labour market and society more generally.

In order to tackle this complexity, a number of meso-level subregional partnerships have emerged which provide some degree of coordination. There is a growing literature on reviewing 'good practice' in dealing with local impacts of migration, designed to help local authorities, Local Strategic Partnerships and other intermediaries to respond positively and proactively to migration (for example, see Audit Commission, 2007; IDeA, 2007; Commission on Integration and Cohesion, 2007; Institute of Community Cohesion; 2007). Often the main emphasis of these more recent 'good-practice' initiatives/guides is on migration from Central and Eastern Europe, focusing on support for migrants and support for the settled community in coming to terms with change, but much of the content is relevant more widely. The overall aim is to provide a more strategic and structured approach to activities at local, regional and national levels, and to develop a more coherent information base on activities.

CONCLUSION

Concerns about labour market integration of immigrants remain at the forefront of the policy agenda in the UK. As the focus of the immigration debate has shifted from refugees to economic migrants, so too has the emphasis swung from a primary focus on 'participation' to incorporate concerns about 'advancement' in the labour market. This is in keeping with an emphasis of policy on better utilisation of skills and progression in employment (DWP and DIUS, 2007). It also chimes with the desire on the part of the UK government to 'manage' migration for economic benefit, by meeting labour market requirements (i.e. filling labour shortages and addressing skills gaps).

It is clear from the discussion in preceding sections of this chapter that the main emphasis to date has been on supply-side initiatives to facilitate labour market integration. The demand side has received much less attention. Yet the behaviour and perceptions of employers and their recruitment and retention policies are central to understanding the experience of immigrants in the labour market. While some employers are keen to recruit migrants to fill key skill shortages, what others want from economic migrants is an easy-to-hire, easy-to-fire, reliable and easy-to-retain workforce to fulfil current production/service requirements at relatively low skill levels where typically they have experienced labour shortages (Anderson et al., 2006; Dench et al., 2006).

Moreover, increasingly concerns are not solely about how refugees and migrant workers fare in the labour market, but also about the impacts on UK nationals. While there are clear economic 'gains' for migration at the macro level (Home Office and DWP, 2007), the 'pains' tend to be experienced at the 'micro' level. Hence the need to ensure that both migrants and UK nationals are equipped to, and have the opportunity to, contribute to economic success.

12 Looking for Work
Exploring the Job Search Methods of Recent Refugees and Migrants

Sonia McKay

In the UK, codes of practice, established by government-appointed commissions, advise on 'appropriate' and nondiscriminatory recruitment procedures. These codes, while not legally binding, are an important soft law mechanism and employers wishing to avoid discrimination claims are aware of the need to ensure that their selection and recruitment practices meet the standards set in the codes, which include the recommendation not to use word-of-mouth recruitment.[1] Consequently, most large UK employers operate formal procedures, requiring of job applicants that they apply in writing, completing an application form and addressing specified criteria related to the requirements of the post. This chapter, by focusing on the job search methods of refugees and recent migrants, explores why, despite government policy to encourage transparency in recruitment through support for formal job search procedures, many recent immigrants show a preference for informal job search mechanisms. The chapter suggests that this may well be because informal mechanisms are more likely to result in job offers. Accordingly, the chapter deconstructs informal job search methods, to demonstrate that they are as complex and as structured as formal methods. The chapter draws on two research projects, one focusing on refugees and the other on recent migrants, in the course of which around 280 semistructured in-depth, face-to-face interviews were undertaken.[2]

HOW INFORMAL JOB SEARCH METHODS FUNCTION

Much of the literature on migrants and refugees notes that informal methods of job search, with reliance on word of mouth and on the recommendations of family or friends, is commonplace (Cox and Watts, 2002; Reyneri, 1998). For example, Mottura (2002)[3] also found that the most common job search mechanisms for migrant workers in Italy were through migrant networks. We wanted to investigate whether over time the type of job search used changes and whether it might be the case that those who have been in the UK longer abandon word of mouth as a primarily tool of job search, focusing on other methods which might deliver better employment outcomes, since informal methods in general give access only to a narrow range of jobs (Baum et al., 2007; Pollert and Wright, 2006). The refugee project has been particularly useful in exploring this question, as it was longitudinal in design, giving researchers the opportunity of interviewing the same group of refugees on a number of occasions, spread over thirty months, and allowing us to explore how their situation had changed over time, while also giving space to participants to reflect on their cumulative experiences of looking for work. We found that the overwhelming majority research participants, all of whom had the legal right to work in the UK without restriction, had initially only used word of mouth in looking for work, relying on contacts they had established within their own communities. Seven in ten respondents indicated this to be the case; only one in ten had used the job centre service to look for work. And just one in ten had themselves initially found work, without community assistance. Participants in the migrant research responded somewhat differently, in that they were much more likely to have sought out their early jobs through the services of employment agencies. The reasons for these differences are related to the presence (or absence) of social networks that might be positioned to assist individuals in obtaining work. In the case of the refugee participants, generally they had joined existing established communities and lived in the areas occupied by these communities. They were able to share information about available work, and in some cases these established networks could even assist applicants to get jobs. Migrants, in contrast, were more likely to begin to use informal methods of accessing work once a community of sufficient size had been established that would allow them to circulate information about available jobs. But this could take a considerable amount of time, dependent on where the emerging migrant community was geographically located. Until that point in time they tended to use employment agencies to find work. The relationship between the existence of well-established communities and ways of accessing work can be confirmed by looking at how geographical location impacted upon choice of job search methods. Among those whom we interviewed, those migrants living in London, where there were already established migrant communities, were most likely to have accessed employment through word of mouth. In contrast, this was the least likely route into employment for those located in newer regions of migration, like the East of England and the South West, where work was more likely to be accessed through employment agencies or labour providers. Word of mouth was also associated with status vulnerability. Those with no authority to work, such as asylum seekers and those migrants who were undocumented, were more likely to have used word of mouth to find work. Equally, the lower paid a job was, the more likely it was to have been accessed through word of mouth. In the migrant study it was the cleaning sector, which paid the lowest wages, where jobs were more likely to have been acquired through word of mouth (McKay et al., 2006). Entry routes into employment may therefore be dependent on the type of jobs being sought. For example, in the migrant study, those in healthcare were more likely to access their jobs from abroad and thus through agencies, although some employers in the processing and packaging and in the hotels and catering sectors were also starting to recruit directly from abroad. Agriculture employers were also more likely to have used agencies to recruit through the then existing Seasonal Agricultural Workers' Scheme (SAWS) (McKay et al., 2006).

The question that then occurs is why it is that informal methods were sustained even once individuals had been in the UK for a considerable (in relation to years rather than months) period of time, particularly as jobs accessed through informal networks can perpetuate unskilled and second-ary-sector employment (Bloch, 2008). Some commentators suggest that it is due to familiarity where the use of informal routes into employment, and, in particular, word of mouth or recommendation, is an established form of job entry into some labour markets and indeed may be the principal form of entry. By contrast, formal methods may be perceived as hostile, as placing applicants 'on trial' and, as one research participant, a doctor and a refugee from the Indian subcontinent, noted, for formal job searches individuals had to be what he referred to as 'thick skinned':

> You have to sell yourself. And—you know—in many cultures . . . people don't usually go and say, I've done this. Because they say OK, I should be modest–you know—they will like me more demure and that stuff, you know. In England, you have to say OK, I've done this, I've done that. I'll do this, I'm capable of doing that. I've done this course.

The concept of having to 'talk oneself up', to boast about one's abilities, does not sit easily in those communities where individuals have more usu-ally relied on third parties to make their case, specifically to avoid being perceived as boastful or arrogant. Recruitment methods, which require of applicants that they respond directly to questions like 'Why do you think you are right for this job?' 'What skills can you bring to the job?' or indeed 'Why do you think you are the best candidate?' present real difficulties for those who would regard responding positively to these types of questions as indicating a boastfulness or conceit.

However, this chapter suggests that, without underestimating the impor-tance cultural norms, the reasons why work is more likely to be accessed through informal channels are more complex than just as a result of cultural or historical preferences. We found that even where respondents were in a position to exercise a choice over job search methods, many still indicated a preference for word of mouth. The reasons for this became apparent once we had analysed our interview data and had found that those who had used informal job search methods had been more successful in finding work. If the individual's primary concern was to get paid employment and if this was more pressing than a desire to get work appropriate to their qualifications— in other words where there was a financial imperative to work—word of mouth represented a very successful method of job access. This was particularly the case for refugees who had forcibly been kept out of the labour market while their claim for asylum had been considered and who therefore were likely to have, during that lengthy period, acquired debts. One research participant, who had been in the UK for five years, and who had looked for work through the job centres and on the Internet, spoke of how, despite the availability of these methods, her successes in obtaining employment had all come through relatives and friends who had provided introductions into jobs. For her this was the best way to get work:

> For me giving the information to my friends that I am looking for a job worked better than newspapers, Internet, any other means because it was quicker and it was answered quickly from the firms and I knew about the firm because my friends told me. I was able to answer the questions very [well].

In addition, our research also suggests that just because a job search method may appear 'informal' does not imply that it occurs in an environment devoid of planning.

Indeed, accessing work through word of mouth turned out to be a more intricate method of job search than it outwardly appeared. Word of mouth is not only a method of communicating information about jobs; it involves complex assessments of the value of the intermediary and their ability to act successfully on the individual's behalf. It involves a conscious selection of a specific intermediary and for specified reasons. Thus, informal job search methods can be characterised by following typology shown in Figure 12.1.

Thus, sourcing work through intermediaries involves a highly complex analysis of the advantages or otherwise of the method on offer. An example of *reciprocal* is the case of a dentist from the Middle East who had been in the UK for five years at the date of interview, and who still expressed a preference for jobs obtained informally. This he put down in part to cultural differences and the way that jobs were obtained in the Middle East, compared to the British system that was heavily reliant on paperwork. However, for him the principal advantage of using informal contacts was that he obtained a speedy answer to his request for work. Sending formal application forms often resulted in no reply or in not being short-listed. But he also emphasised that it was not just a matter of obtaining information about jobs from any intermediary; the process involved a careful selection choosing a 'trustworthy friend' to act as the intermediary, someone who had the respect of others. This notion of reciprocity—assessing the worth of the intermediary to the individual job searcher and the worth of the job searcher to the intermediary—emerged in a number of other interviews, particularly in the refugee project, and was not dependent on the level of job being sought. Thus, the need to ensure that the right intermediary acted on the individual's behalf was just as likely to be expressed in the case of those with minimum educational qualifications,

(a) *Reciprocal or trust action* involving an assessment of:
  • The standing of the intermediary in the community;
  • The individual's standing in relation to the intermediary; or
  • The intermediary's standing in relation to the potential employer;
(b) *Recommendation*—a process involving the evaluation of the intermediary's employment and accepting their description of it being good work;
(c) *Empathy*—an acknowledgement of the intermediary's understanding of the individual's qualifications and employment needs;
(d) *Insider*—where intermediaries located within the workplace enable the individual to present her/himself within the workplace as someone who is connected and who thus is not an 'outsider';
(e) *Guidance*—the formal acceptance of direction from friends/family towards work opportunities; or
(f) *Acquaintance*—a means of using intermediaries to acquire knowledge about the jobs market.

Figure 12.1 A Typology for Word of Mouth

seeking low-skilled work, as it was with those seeking to enter professional posts, such as that of the dentist cited earlier. In the following case, Abdul, a young male from East Africa, had been in the UK for two years. Educated to secondary school level and seeking a job as a sports coach, he similarly emphasised how important it was to carefully select an intermediary with appropriate standing in relation to the job being sought.

> Because sometimes it depends on somebody's experience and reputation you know, that's what I'm looking for. You know if you are working in some place, some place or some factory or anywhere and he's a good man and the people are looking for people to work with him so he can introduce you to his boss and then [if] his reputation is good, that's why they will accept you. Because your friend is good.

Afshin, from the Middle East, educated to tertiary level, still used intermediaries as his preferred route into employment, even though the jobs he had obtained were in illegal leaflet distribution, security and hotels; assessing the value of his intermediaries on the basis of:

> First the connections, second their ability, third their capability, fourth ah, their trustful, how much they trust him or he trust the others um, how ah, important you for him to do this thing for you or not. . . . the importance of your relationship, your personality, your prestige.

Sarabjit, a doctor from the Indian subcontinent, had been in the UK for three years and similarly expressed the view that he would obtain employment 'only if I know any person is belonging to—for example to my community, will support me, for example the person should be a consultant then I can get a job'.

Intermediaries may also be used where the individual assesses the intermediary's employment position favourably, characterised here as '*recommended*'. In these cases the intermediary has either offered favourable descriptions of her/his employment situation or has otherwise conveyed such a message that in turn encourages the individual to rely on that intermediary for support into employment. Jamal, from the Middle East, had been hoping to work as a bus driver but, unable to obtain this employment, had scaled down his sights and was aiming at work as a self-employed van driver, because he had a friend who was successfully engaged in that type of work:

> That's why I'm looking for how to get in that; especially there is some kind of delivery which they call owner-driver. You become self employed and you are buying your van and you are delivering stuff for that company . . . So one of my friends is doing this job, he's quite happy so—I am trying to get into this job. Yeah, it's good.

Mary had been in the UK for eight years but had never used formal methods, like the job centres or job adverts, to access work. Although she aspired eventually to qualify as a nurse or social worker, in the meantime she accesses factory work where it is recommended by others 'My friend told me the work is somewhere in a factory, he said it is a very good factory and they take everybody, so we went there'. Intermediaries may also be used in cases where it is believed that the intermediary has '*empathy*' for the job searcher's situation and a good understanding of her/his needs in relation to employment. This is most likely to be the case where the job searcher has additional commitments, such as studying or childcare, and also where the intermediary is believed to have a good understanding of the individual's abilities. In these cases direct approaches to employers or direct responses to notices for jobs are viewed as methods where talents will not be recognised. An intermediary thus is seen as being in a better position to search out suitable work, understanding the individual's needs and abilities. Nahid, a 38-year-old woman from Iran, educated to tertiary level, explained why an intermediary could act on her behalf:

> Quite informal—that is true it is quite informal but when a friend—for example [hears about] a job application she or he knows me and exactly knows what is my desire and what is it exactly I am looking for and these kind of things. So that is why normally, when I get these kind of e-mails and these kind of information it is quite close to what I want. And it is quite close to my abilities and my previous work-experience. But when you go to some sort of formal job search or—you mentioned

about job centre these kind of things, well they don't know exactly my language abilities; they don't know exactly my previous experiences. Well in general they can give you some information but in particular no they don't know [me] so that is why certainly I think . . . informal job-search is quite effective.

A respondent seeking part-time work while continuing his studies also per-ceived this type of work as something that could not easily be obtained without the intervention of an intermediary. In his view, if he were to apply formally for jobs advertised through newspapers or the job centre he would 'have to bow to the employer what he will offer me' and the flexibility he desired would not be there. However, this choice of route to employment also forced him only into work that was located within his own community:

I am just asking friends to be honest. Especially if you are willing to work part-time, you need someone who can understand your position. I can't go to any employer only part-time because if I want to go for a course or something they will not understand but if they are people who understand my position, OK you can go these two days, don't come on these two days but if you are going to cover another two days, that won't happen in any business, only in a business that they can really understand what is going on. . . . That is why I kept that to mostly to my own community because they can understand.

The research also suggests that intermediaries will be used where individu-als desire to be seen as *'insiders'* who would be 'accepted' in the workplace. Thus, the intermediary can facilitate entry into the workplace as well as assisting in obtaining employment. Maisoon, a qualified teacher from the Middle East, had been in the UK for more than five years, yet in moving into her first UK teaching job she felt encouraged because she already knew the head teacher and felt that she could take up this job because in the cir-cumstances 'you don't feel like a stranger'. Seeking out work through inter-mediaries also means acquiescing to the *guidance* of others and accepting their advice and recommendations in relation to work opportunities. When Kibar first had permission to work she was limited in the jobs she felt she could apply for, due to her limited skills in English. As a result she ended up working in a textiles factory: 'my friend said to me to go there and work. No, no straight into textile. Because I didn't know anything else and I mean I also couldn't speak the language. I didn't have an option to go somewhere else. This was the most sensible thing for me to do'. Nira, from Sri Lanka, similarly described how friends and neighbours directed her towards a par-ticular type of work:

[I got the job in] Co A because my brother worked there Saturday and Sunday [as a] manager, so when I came here he said you can work part-time there, we have a vacancy there so I worked there. [That] was [the] first application. Then second [CoB] one of the ladies travelled by bus, she normally talked to me, one of the Indian ladies, she told me we have a vacancy in Co B, if you apply, go and ask the manager for the application form.

Finally, intermediaries may also be used to obtain information and *knowledge* about the labour market. Karzan used his friends and family to 'get information'. He has: 'Friends who work in factories and stuff and I get information from them, not because I want to work in the factory, but just for my own knowledge'.

Even here the calculations being made are quite complex. Accessing work through first gaining knowledge depends not solely on the ability to use intermediaries but also on who the employer is, with a recognition that workplaces of high-trust management are better for accessing work through word of mouth. Masood had been in the UK for four years at the time of interview and expressed this view well:

Because [in] some places [you] cannot find jobs from friends. Because the management of that place they trust their workers. They tell workers to find people for them. But some of the companies they do not trust the people who are working for them. They trust the agencies. Or the jobcentres. Or some places like that. So it depends. I mean I used to work there. I took three of my friends to there. And they are still there. They are still working yeah; I took them to there. And I asked the operation manager. And they accepted them because of me.

However, in 'low-trust' firms—firms where relationships between employers and workers are poor—word of mouth was unlikely to be used, as it was acknowledged that here an internal intermediary would not be able to facilitate an offer of employment. In such cases some interviewees assessed agencies as a better route into such workplaces.

As already indicated, word of mouth was the most likely route that refugee interviewees used, not just in accessing their first employment, but in accessing employment at any stage of their period in the UK. For migrant workers, word of mouth was more likely for second and subsequent jobs. And although it was the case that word of mouth was supplemented by other forms of job access, particularly the longer someone had been in the UK, it remained a method that interviewees consistently viewed as successful, where success was measured in the ability to secure employment.

The research we conducted has also included face-to-face in-depth interviews with more than eighty employers (including employment agencies). In conducting these interviews we also found that there was a similar preference from employers for word-of-mouth recruitment, particularly for low-skilled work and also for initial employment. A transport employer in South Wales had introduced a 'recruit-a-friend' scheme. This was seen as a successful way of accessing additional labour. Others spoke of established workers passing on information to new arrivals. Around half of the employers interviewed in the migrant-worker research had at one time used employment agencies, although a number had switched from agency employment to word of mouth, as the migrant workforce increased: 'The agencies are, almost made redundant by the fact, what I call a second generation now coming in, friends and relatives are turning up on the doorstep' (Food employer, South West).

One employer spoke of no longer having a need to advertise, as applicants 'are constantly coming looking for work without the need for us to make any advertisements in magazines or papers or whatever, people just turn up' (Catering employer, London). This company also maintained a database of migrant workers who had requested employment, and as jobs came up would contact them and, provided they matched the necessary qualifications for the post, would be taken on. But there were also cases where employers had shifted from using employment agencies due to concerns over the employment terms and conditions that agencies offered to the workers they supplied. Some were also concerned over the instability of the agency workforce and the consequent implications for health and safety and for this reason were turning to direct employment.

FORMAL JOB SEARCH METHODS

Using the Job Centre

Jobcentre Plus is a government-funded agency to assist individuals into employment. Employers can register job vacancies with Jobcentre Plus and individuals can use the expertise of the centres to find work. Importantly, individuals who are unemployed must register with the job centre and demonstrate that they are actively pursing work. However, the available evidence suggests that the job centres are underutilised by refugees and recent migrants (Bloch, 2002; Dhudwar, 2004; McCabe et al., 2006; Phillimore and Goodson, 2006). In our research on refugees we found that both men and women who had used job centres in looking for work often spoke negatively about their value in relation to finding jobs. Some participants had never used the job centres, but of those who had, it had not necessarily led to employment, and they spoke of finding work through other means, with word of mouth being assessed generally as the most successful route into employment. Older interviewee participants were more likely to have used job centres, as were those who had benefited from a tertiary education. However, it was precisely this group of older, more formally educated participants that also was more likely to have a negative assessment of the value of job centres, categorising them as not offering access to the type of job they aspired to do. Job centres were seen as forcing professional workers into low-paid work that did not reflect their qualifications. One Iranian participant was clear that if what was being sought was a professional job, the job centre was not the appropriate place to go, stating, 'If you want bread, you are not going to [a] bookshop for bread because you know that shop can't provide what you want'. Indeed, those participants, who at one stage had used job centres, tended to move to other forms of accessing work, the longer they were in the UK. Dissatisfaction with the service provided by job centres was expressed in relation to never hearing back about applications; poor treatment at job centres, with job-centre staff sometimes being perceived as discouraging of individuals' aspirations; and of no-one at the job centre being on hand to assist in job searches. Additionally, there was a view that employers were cautious about applicants who had been referred through the job centres, seeing them as in some way 'unemployable'. Some participants told of being sent to jobs by the job centre, only to find out that the job was not available when they turned up. They also complained that no attempts had been made by the job-centre staff to contact them following the interview, to find out what had happed and perhaps to provide advice on the process. One described the job centres as a 'dictatorship, unhelpful and discriminatory'. Such views were based on perceptions that, in the view of the job-centre staff, refugees and migrants were unqualified and only suitable for low-paid work. Job-centre staff had told Nadia that they could not provide her with information on jobs in her profession. For her the job centre was unsympathetic to her attempts to gain a post in medicine.

> Actually once I talked to the Job Centre and they said they will not provide us with a job which is related to my qualification. They said if they provide me, they provide me job like cleaner you know, working in factory, packing, these things and I said look I'm a qualified doctor. And they said we don't care who you are in the job centre we just can you know offer you this kind of job.

Another participant commented:

> They [the job centre] tried to humiliate me and that was very frustrating for me every two weeks go there sign in and to explain what you're doing or what you're not and they have this impression that we don't want to work. And they think we want just for the benefit. They have not enough knowledge they have not enough information about us.

Those who were in receipt of state benefits had no alternative but to register with the job centre or lose their benefits. But they too spoke of frustrations and of job-centre staff being inexperienced and unable to offer the support needed and of assuming always that their problems in relation to getting work were related to limitations in spoken or written English. One participant emphasised that whatever his experiences or qualifications, the job-centre staff would always see him as lacking in some essential attribute, be it language, training and education, experience or skill. And even where he could demonstrate that he had all of these attributes, the job-centre staff would then refer to additional attributes that he happened not to possess. For some participants, the job centre just did not have the necessary information or understanding of the difficulties that they were experiencing in trying to get suitable work. Josephine, a qualified nurse, had been in the UK for two years, trying to get her qualifications recognised, and in the meantime was told to apply for a cleaning job in the transport sector. When she refused, her benefit was stopped and her arguments that she was in the process of registering for nursing were ignored. As she noted:

> Forcing me to do the cleaning job, which, which is not even going to solve my problem. Because my problem was to get the job, be fully independent, be able to pay my own rent, to feed myself and my children, to be able to support myself.

Another frequent criticism was that when applications for work were submitted through the job centres, participants either never heard back from the prospective employers or were told that the jobs were no longer available. Although interviewees were usually reluctant to express it as such, the background that they presented in telling their story in the interview raised issues of race or nationality discrimination. Afshin, recounting his experiences of going for jobs that were always already filled, spoke of occasions of finding out weeks later that the same jobs were still being advertised. Those relatively new to the UK labour market expressed surprise and concern, when they had submitted written applications for jobs, at later discovering that there had been an internal candidate and that there really was little chance of the job being offered externally. Tades was looking to work as an ESOL teacher and spoke of applying for jobs where there were internal candidates. In his view the jobs were advertised 'just to make it legal, they advertised it [the job]. I met those people [the successful applicants] they told me they were working there, so in that particular situation I accept'. In this case it was not that Tades objected to an internal candidate with relevant experience being selected over him; it was the pretence that this was a genuinely open recruitment that caused frustration, and this was a pretence that the job centres were seen as conspiring in maintaining. Several participants additionally referred to additional difficulties they had encountered in using the job centres due to the fact that at the centres there is a their heavy reliance on telephones. Individuals are often asked to communicate both with job-centre staff and with potential employers by telephone and job-centre staff were often not physically present and only communicated with them by phone. Those who were less confident in their English-language skills found this particularly difficult.

While those with high human capital generally felt that the job centres could not assist them into appropriate work, those with low human capital generally found the official job-centre service more useful. For example, a participant, looking for bus-driving work, had found the job centres helpful and useful as he moved around the country taking up different jobs. For him the job centres were well equipped to provide access to these types of jobs. Those with positive experiences of job centres also based their assessment on the fact that the job centres were free to use, could help with CVs or job application forms and importantly, that a failure to register with the job centre might lead to loss of benefit. Positive attributes such as these were emphasised by Mardin, a refugee from the Middle East, who had been residing in the UK for just under five years. He pointed out that the advantage of the job centre was that it could be visited every day and was free. Other regular job search methods, like checking newspaper job adverts or the Internet, cost money.

Overall, although there were both positive and negative assessments of the value of job centres in relation to job searches, in general it can be said that when it came to accessing employment, participants were overwhelming in their view that the best way of getting work was through informal recruitment mechanisms. What we did note was a pattern, of individuals starting to use job centres once they had been in the UK for a couple of years and then, after a short time, this method of job search gave way to other approaches, with individuals increasingly reliant on their own job-seeking abilities, conforming to the typology below.

Word of mouth: (moves on to →)
Job centre: (moves on to →)
Employment agency/newspapers/Internet

Thus, individuals initially relied on word of mouth; then they moved on to more formal methods of work access, initially testing the services offered by job centres. However, once they had reached a stage when they saw themselves as better established in the UK, having already done some paid work, they were more likely to rely on jobs that they accessed directly or through employment agencies, newspapers or the Internet.

ACCESSING JOBS THROUGH EMPLOYMENT AGENCIES

Our research points to a pattern of an eventual greater reliance on employment agencies for accessing work, particularly for those groups that had initially relied on word of mouth. This may result from a realisation of the type of jobs that were readily available and that were most easily accessed were jobs that were advertised through employment agencies. With rare exceptions, we found that employment agencies were used to access low-skilled, low-paid work. In many cases this was also the type of work that could equally have been accessed through the job centres or indeed, in the early stages of an individual's presence in the UK, might have been accessed by word of mouth. The reasons why participants chose to use employment agencies were associated with a view that if the only jobs that could be obtained through the job centres were low skilled, then it was actually better to source these through agencies. As one participant pointed out, the process for getting work through the job centre was lengthy, whereas the agency could direct the applicant to unskilled jobs right away. Indeed, for this kind of low-skilled manual labour accessing jobs through the job centre, by contrast, was 'a waste of time'.

> I think agency is better than Job Centre. And sometimes they say directly we haven't any job or we have job, would you like t—if you go to the Job Centre you can try some kind of job, apply for some kind of job for yourself, checking the information. After they try to call the company and you can get application form from the employer. You can send back to them but you probably won't get a response letter. So I think better way is to go to agency directly.

Karwan, a refugee from Iraq, similarly noted that 'agencies are known by their competence and they can find you a position'. Other reasons given for using agencies included that employers tended to go through agencies to test workers first before deciding to offer employment directly or because the criteria for appointment might be less rigorous, as agencies supplied labour in situations where employers had an urgent need for it. Thus, those who considered their English-language skills to be limited were more likely to believe that the agencies gave them better access to work. Not only could the agencies offer jobs more speedily but also individuals did not internalise the jobs accessed through agencies as in any way permanent or a definer of their future occupational status, in the way that they did when considering jobs accessed through a government institution, like Jobcentre Plus. Thus it could be that workers who aspired eventually to obtain jobs that were related to their own qualifications or experience were more comfortable if accessing lower-graded jobs through agencies, precisely because the nature of the agency work they were doing was clearly associated with a temporary situation and did not identify the individual permanently as associated with that occupation. Thus a dentist could work as an agency cleaner, while still maintaining internally his identity as a professional worker who was merely temporally in a different role; a teacher could work as a textile worker, still defining herself as a teaching professional. Where the same kinds of low-skilled work had been accessed through a government agency, it was perceived as assigning to the individuals concerned a more permanent work identity. It may also be that participants using agencies saw themselves as still leaving open a path to alternative more suitable employment when the opportunity arose.

Of course, not all of the views expressed on employment agencies were positive. Some spoke of agencies never contacting them or of jobs going to local workers. Interviewees also made the point that a number of employment agencies no longer had available the range of jobs that had been on offer a few years earlier. Recent migrant workers from the new EU states were filling jobs that previously would have been done by refugees, according to one interviewee, whereas, in the view of another, jobs were being offered to minority ethnic workers who were already settled in the UK or who were the children of an earlier generation of settled migrants.

CONCLUSION

This chapter has focused on how immigrants new to the labour market access first and subsequent jobs. It has found a preference for informal methods of job search, in particular the use of friends and community networks, for those new to the labour market, with, in the case of refugees, a specific preference for this form, where individuals find themselves geographically located within areas with an established community presence. Similarly, in relation to economic migrants, informal job search methods are used where the individual is geographically located in an area with a preexisting community. However, where such communities do not exist, the preference still is to stay outside the established and state-sponsored Jobcentre Plus, with employment sourced through an agency or labour provider often being preferred. The chapter has argued that explanations for this lie firstly in the need to acquire work quickly. Family and social networks, together with employment agencies and labour providers, are viewed as best able to do this, by opening routes to the sort of work that is easily available, even if it is not commensurate with skills and qualifications. Moreover, jobs accessed outside the formal state system can be distinguished from 'real' jobs and can be taken on as temporary solutions to an immediate problem (of finances), rather than as the reconstruction of a new work identity in lower-skilled employment. What can be concluded from this analysis? First, there is a strong evidence base for viewing the state-sponsored job search models as inappropriate or at the very least not able to respond to the needs of recent immigrants. The fact that so many refugees and economic migrants appear to have 'voted with their feet' and used informal job search methods testifies to their firm belief in the inability of the state to assist them into work, even when they are in the UK with a legal right to work. But furthermore, the evidence presented in the chapter points to a need to rethink notions of equality and fair play in relation to recruitment. The principle that all applications for work should be similarly processed and that jobs should only be given on merit in reality hides what too frequently occurs. As many interviewees noted, just by advertising a job and proclaiming that it is open equally to all candidates does not deal with what in practice happens. Internal candidates are often favoured; job specifications may be drawn up with particular individuals in mind; and the obligation to complete and submit endless numbers of job applications,[4] to retain entitlement to state benefits, may indeed amount to a wasted effort which can only lead to demoralisation among those seeking work. It is for these reasons that recent immigrants take a clear and rational decision to construct their own routes into work. It would be better if this was openly acknowledged to be the case. Employers need to be more honest and admit that some jobs are not equally available to all while the state needs to reexamine its job search services, to reconstruct them in a way that would enable it to respond appropriately to the needs of recent immigrants. This may include a recognition of the value of word of mouth for new entrants into the labour force, together with some clearer advice to employers on how to ensure not just that their recruitment procedures are transparent and that they apply equally, but also that, in terms of outcomes, they deliver a heterogeneous labour force which includes recently arrived refugees and economic migrants.

NOTES

1. Para 4.11a of the statutory code says 'Employers should avoid recruitment, solely or in the first instance, on the basis of recommendations by existing staff, particularly when the workforce is wholly or predominantly from one racial group.
2. The first was an ESF-funded project on comparing the labour market experiences of refugees and ethnic minorities. The second was a UK government agency–funded project on migrant worker health and safety. (See McKay et al., 2006 and McKay et al., 2006a.)
3. See Chapter 6 of this book.
4. In the refugee project, some interviewees who were still unemployed had submitted several hundreds of application forms, without ever receiving a favourable response.

13 Employability Initiatives for Refugees in the EU
Building on Good Practice

Jenny Phillimore

This chapter draws upon research undertaken to examine the types of employability initiatives being delivered across the UK and Europe. Using a combination of qualitative and quantitative techniques, the research explores the types of initiative on offer, the range of organisations that are providing employability initiatives, barriers to provision and good practice within existing provision. The research finds that a combination of well-structured work experience, combined with positive employer participation and on-site accreditation of prior learning, is necessary for maximising refugees' chances of employment. It highlights the difficulties that organisations in the UK experience in delivering employability support, given the lack of core funding for these activities and the pressures they experience dealing with refugees' other concerns, including accessing housing and appropriate English-language training.

Until recently most refugees arrived in the UK under the auspices of refugee programmes set up across Europe or beyond as a result of actions relating to particular global political concerns, for example, the expulsion of Asians from Uganda, the Vietnamese boat people and the so-called Bosnian crisis (Sales, 2002; Kuepper et al., 1975). It was not until relatively recently that the numbers of asylum applicants arriving spontaneously in Europe began to increase to the extent that successive governments moved to develop policy initiatives to address both the increase in numbers and the impact that this increase was said to have on services around receiving areas. The situation in the UK reflects that in much of the rest of Europe. The past ten years have seen the arrival of large numbers of people seeking asylum with figures reaching an all-time high of 110,000 applications in 2002 (Home Office, 2003). In response to the pressure asylum seekers had been placing on support services in London and Southern England, the National Asylum Support Service (NASS) was established in 1999 to coordinate and fund the dispersal of asylum seekers around the country. Concerns have been expressed by refugee and civil-rights groups that moving asylum seekers away from the well-developed networks of friends, family, legal and other support services in London would hamper their ability to settle and locate employment (Zetter and Pearl, 2000).

Changes in asylum policy have been accompanied by an increasing interest in integration. *Full and Equal Citizens* (Home Office, 2000) and *Secure Borders Safe Haven* (Home, Office 2002) developed some ideas for the integration of 'recognised' refugees, the promotion of 'citizenship' and for the establishment of a National Refugee Integration Forum to explore the different dimensions of integration and examine how they might be implemented in policy terms. More recently, the publication of two further papers, Ager and Strang's (2004) report setting out an Indicators of Integration framework and *Integration Matters*, the Home Office's (2005) most recent strategy for refugee integration, proposes a set of actions aimed at taking the strategy forward and then presents some indicators, based on Ager and Strang's framework, for measuring success. Both place considerable emphasis on the key role of employment in facilitating the integration of new migrants. Employment has long been considered one of the key factors in the integration of refugees into society (Mesthenos and Ioannidi, 2002; Robinson, 1999). There has been a great deal of research exploring the barriers that refugees face when attempting to gain access to the labour market (for example, Aldridge and Waddington, 2001; Bloch, 2002). Far fewer studies have examined approaches to overcoming barriers to employment.

This chapter draws upon research undertaken to examine the types of employability initiative being delivered across the UK and Europe. Using a variety of research techniques, the research explores the range of organisations that are providing employability initiatives, the services they offer, barriers to provision, and good practice within existing provision. The chapter looks at the differences between employability initiatives in the UK and Northern Europe. The chapter asks what might be learned from the Northern European approach to enhancing refugee employability and how such an approach might be utilised in the UK to help to overcome the key barriers to employability.

REFUGEES AND EMPLOYMENT

There has been much debate about the meaning of the term 'integration' (see Castles et al., 2003) but considerable agreement that integration is a complex, multifaceted process that operates in a range of dimensions from the social and political to the functional (see Zetter et al., 2002). From a policy and service-provision perspective, most emphasis is placed upon the functional dimension. Within this, employment is considered to be the single most important factor in securing the integration of migrants into society. Coussey (2000), Bloch (1999), Robinson (1998), Srinivasan (1994), Joly (1996), and Knox (1997) all argue employment is the key to successful integration. Employment enables interaction with hosts, increases opportunities for learning English, the opportunity to build a future and to regain confidence and esteem whilst economic independence makes adjustment to society easier. Indeed, improving access to employment is a key policy goal in the UK where the government has focused on paid employment as the path to social inclusion. Policies have been aimed at creating initiatives under the New Deal programme specifically for historically excluded groups including ethnic minorities, single parents, young people and the long-term unemployed.

Research by Bloch (1999, 2000) suggests that those refugees who are working adjust more easily to the host society than those who are unemployed. Employment has been found to be crucial to the psychological well-being of ethnic-minority migrants, whilst unemployment has a significant adverse effect. In particular, women who were able to work experienced significantly enhanced psychosocial well-being compared to the unemployed (Shields and Wheatley Price, 2003). Indeed, it has been argued that

> for a refugee, who has been powerlessly dependent on the benevolence of the receiving country, the psychological value of obtaining a job will be greater even than for an indigenous worker . . . a job will often provide a context where the refugee can improve language skills and come to terms with the social environment of the receiving country. (Robinson, 1998: 155)

Yet evidence exists that inability to locate work is the single most significant barrier to successful integration of refugees into British society (Feeney, 2000). This is coupled with the harder-to-document problem of underemployment as refugees struggle to locate employment commensurate with their skills, and the process of integration is often associated with downward professional mobility (Sargeant and Forna, 2001; Mestheneos and Ioannidi, 2002). Research findings suggest that gaining employment is the main priority of those who have recently been awarded refugee status (Phillimore et al., 2003; Bloch, 2002). Between 80 and 96 per cent of asylum seekers want to work once they are legally entitled to employment (Aldridge and Waddington, 2001; European Commission, 2001; Home Office, 2001; mbA 1999; Scottish Refugee Council, 2001; Phillimore and Goodson, 2001; Phillimore et al., 2003).

Yet evidence clearly demonstrates that refugees are excluded from the labour market in large numbers. For example, the 2000 Peabody Report focusing on London found that 51 per cent of those who had been in the UK for five to eight years were unemployed (cited in Midlands Refugee Council, 2001). The situation for the most recent cohort of asylum seekers appears to be far worse. Bloch found that only 29 per cent of refugees were working at the time of her survey (2002) and that they tended to be employed in low-paid, temporary and unskilled work. Other authors have found rates lying between 60 per cent (Sargeant and Forna, 2001; Feeney, 2000) and 81 per cent (Walters and Egan, 1996). The situation appears particularly serious for refugee women, with mbA (1999) data showing that only 7 per cent of refugee women were employed and with a report on refugee women in London finding that they had high levels of skills but were amongst the most excluded from the labour market because they lacked access to the conventional support systems available to women in the UK (Dumper, 2002). Other research has found that migrants generally fair worse than the UK born, in terms of locating work, with those from industrialised countries doing better than those from poorer countries (Haque, 2003; Bloch, 2002). These respondents also stated that work was the main mechanism by which they would be able to 'fit in' to society in the UK. These findings were supported by Bloch (2000), whose interviews exploring ways of improving refugees' quality of life in North East London revealed that refugees placed employment at the top of their list of priorities. Employment disadvantages existed at all qualification and skills levels, but the situation was more favourable for those from English-speaking backgrounds.

Little data exist to demonstrate the extent to which refugees are underemployed, although the research in Coventry found that 66 per cent of respondents who were employed before arriving in the UK had worked in skilled or professional positions. Of those now working, the majority were employed in elementary occupations (58%) and processing and machine operating (29%). None of the managers or professionals surveyed had reentered skilled work (Phillimore et al., 2003). Such high rates of unemployment are remarkable given a national average in the UK of 5.2 per cent in 2002 (Social Trends, 2002), the high levels of skills possessed by refugees and the skills shortage experienced across Europe at that time. Refugee unemployment rates are more than ten times the national average and underemployment is the norm (Bloch, 1999, 2000; Aldridge and Waddington, 2001; Phillimore and Goodson, 2001). Whilst there are regional and subregional differences in labour markets, in the present economic climate the issue is not a lack of skilled job opportunities. Menz (2002) identifies two types of skills shortages in the EU: high-skilled and seasonal or low-skilled work. At a time of increased globalisation of competition for skills, recruitment of skilled personnel has become a struggle for many organisations in Europe. The report *Skills in England 2004* (DFES, 2005) outlined a range of skills shortages likely in the period leading up to 2010. Key areas of shortage include health professionals and associate professionals, teachers, carers, salespeople, administrators and transport-related occupations, to mention but a few. In 2001 Reed Recruitment, a leading UK employment agency, reported three-quarters of companies trying to cope with skills shortages (Sargeant and Forna, 2001: 9). Although in Europe there has been a move from an internal to an external labour market, based on the work-permit system, the supply of jobs still outstrips the supply of workers to fill them. Refugees form a sizeable and growing community in the UK. Many refugees are skilled and most are highly motivated to work and could fill skills gaps (see Phillimore and Goodson, 2002; mbA, 1999).

Much research has explored the barriers to refugees gaining employment in the UK. These barriers include inability to speak or write English to the standard required by employers, lack of knowledge about UK employment culture or vocational language, lack of UK work experience, UK qualifications and a UK employer reference, and difficulties gaining recognition for overseas skills or qualifications (Bloch, 2002; Dunn and Somerville, 2004; Kempton, 2002; Aldridge et al., 2005). Although we now have a very clear idea of the kinds of barriers refugees face when trying to enter the labour market, we know far less about the ways in which refugees can be supported into work. Refugee unemployment levels in the UK suggest that any initiatives that do exist have been less than successful. This chapter reports the findings of a research programme that was part of an EQUAL-funded project. The aim of the research was to explore European approaches to improving refugee employability, identify innovative practice and develop an employability model that could be piloted in the UK. The chapter reports the findings of this research and primarily looks at what works in employability initiatives. The latter part of the chapter focuses on what can be learned from the Northern European approach to enhancing refugee employability.

METHODOLOGY

The research reported in this chapter was undertaken between January and July 2005 and involved both quantitative and qualitative methods. The research began with the construction of a database of organisations offering employability services to refugees or migrants.[1] The types of services considered included employability advice, arrangement of work placements, employer work, and accreditation and translation of overseas qualifications. We used existing databases provided by refugee or migrant-focused organisations, searched the Internet to identify others, networked with our own and others' EQUAL transnational partners, used a snowballing approach with those identified and used the Web sites of well-known refugee organisations. Eventually we constructed a database of 556 organisations of which 227 were based in the UK and 329 in Europe. The next phase of the research involved designing a questionnaire to help us to identify the types of support that organisations offered. The main areas covered included organisational details, client profile, services provided, general education and training provided, language training, recognising skills and qualifications, work with employers and monitoring. The questionnaire was piloted with our UK and transnational partners and then translated into Dutch, French, German, Greek, Italian and Spanish and administered by post and e-mail where possible. Questionnaires were sent out in different waves depending on when the organisation was identified. All UK nonrespondents were reminded by telephone. Some eighty-five organisations returned their questionnaires by post or e-mail; eighty completed the form over the telephone. Refugee community organisations are acknowledged to be quite transient owing to lack of funds and overreliance on volunteers (Zetter and Pearl, 2000), so we were not surprised to find that 166 were no longer operating. Spanish, French and Scandinavian respondents were contacted by telephone, but elsewhere in Europe reminders were sent by e-mail or post. Response rates were affected by this mixed approach to completion. In the UK we achieved a response rate of 54 per cent, whereas across the EU it averaged at 20 per cent. Questionnaire data were analysed using SPSS.

All organisations providing some kind of employability support including work experience or qualification accreditation services were contacted by telephone to explore the range of services they offered and whether they were willing to host a study visit. We did not contact any organisations offering solely language services or those not serving refugees. In total thirty-seven organisations were telephoned. We chose not to visit nine organisations because their services were not sufficiently relevant to justify further enquiry. Organisations were visited in Norway (4 organisations), Germany (2), the Netherlands (12), London (5), Scotland (2), Leeds (1), Sheffield (1) and Birmingham (1). Two organisations in Norway and one in the Netherlands were included in this part of the study at the suggestion of other organisations from those countries and had not originally been included on our database. Each visit incorporated an in-depth interview with staff working in the relevant areas and where possible interviews with refugees. Issues covered policy context (where overseas), aims, objectives and funding, step-by-step discussion of their employability processes, exploration of relationships with employers, consideration of support available to refugees, programme monitoring, good practice and success stories, discussion around issues of integration and barriers to delivery. We also used the same topic guide to undertake telephone interviews with organisations that were unable to host a visit. These included two in Leicester, one in Wolverhampton, one in London, two in France and one in Luxembourg.

CONSULTATION WITH KEY PLAYERS

In order to explore what kinds of employability model might work in the UK, we consulted with a range of key players. This included interviewing thirty-two refugees about the types of employability they felt where most likely to be attractive and effective, interviewing Jobcentre Plus to see how initiatives could be fitted around benefits regulations, and speaking to employers and their representatives about what kinds of initiative they would be prepared to support. We also held discussions with accreditation bodies such as Open College Network and City and Guilds and colleges about the different ways in which refugees' skills and experience might be accredited. We talked with policymakers at local, regional and national levels about how change might be initiated. Overall, these interviews helped shaped our ideas about what might be achievable in the context of the UK. This chapter focuses upon the findings from the questionnaire and site visits but also features some of the findings from key-player discussions where relevant.

FINDINGS

The Organisations

Responses were received from a total of 165 organisations from fifteen different countries. Some 61.2 per cent of the total responses were from UK organisations (101 organisations), fifteen were from the Netherlands, ten from Austria, ten from Germany, nine from France, three each from Denmark, Italy, Norway and Spain, two each from Czech Republic and Greece and one each from Belgium, Hungary, Ireland and Luxembourg. Half of UK-based organisations (51%) had been established more than ten years. Some 27 per cent had been established less than five years. Many of the refugee organisations based in the English regions had been established post-2000 (see Figure 13.1). Their development represents a burst of establishment activity since the establishment of NASS in 2001. This is perhaps not surprising given the role that RCOs in the UK have adopted. Zetter and Pearl (2000) found that these organisations took the main burden of supporting asylum seekers and refugees to integrate. In addition, the UK's

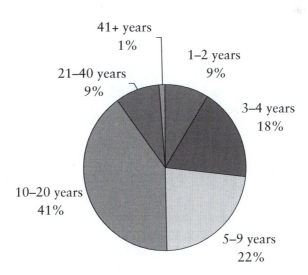

Figure 13.1 Years spent working in the sector (UK organisations only).

Home Office has made some investment into establishing RCOs under the Refugee Community Development Fund. There was a marked difference in the geographical areas served by UK and non-UK organisations. Over half of UK (54%) organisations served only their local area, compared to 6 per cent of non-UK organisations, a trend that may again be attributable to the tendency in the UK to give seed-corn funding to RCOs. The site visits confirmed that RCOs were far less important in service provision outside of the UK, where most employability services were core funded by the state and contracted to municipalities (Norway) or colleges (Netherlands). Thus, overseas organisations tended to operate at a larger scale and work regionally. Organisations in Scandinavia considered the lack of RCOs in their countries to be problematic and wanted to see RCO development occur to complement the services provided by the state and help to make those services more user friendly.

In terms of scale, there were some differences between UK and non-UK organisations. Non-UK organisations often served refugees as part of general services to the whole population whereas UK organisations often specialised in services for refugees. The majority of organisations were small scale, serving less that fifty refugees in 2004, although in the UK 11.2 per cent of the sample served over 1,000 refugees, whereas no non-UK organisation served this number of people. This may reflect differences in approaches to housing refugees. For example, in the UK refugees are concentrated in the dispersal areas where they were housed while their asylum claim was processed (Phillimore, 2005), whereas in Northern Europe refugees are dispersed on a quota basis, after their claim is processed, so tend to be living in a large number of areas in very small numbers. The majority of small (less than 50 clients) organisations in the UK had been established since the introduction of dispersal, suggesting that they were fledgling organisations.

The Services Offered

We also explored the range of services that organisations offered. There were some similarities in the services offered by non-UK organisations and UK organisations, although, outside of the UK, slightly more emphasis was given to employment, youth and women services, and slightly less on training (see Figure 13.2). The service most frequently offered was that of training. Some 65 per cent of organisations provided language training; 50 per cent provided vocational training and work-based learning. The majority of organisations offering work-based learning were based overseas. Some 69 per cent of all organisations provided courses to help migrants prepare for work. Most organisations offered more that one service. The highest proportion of refugee-serving organisations (58.8%) provided one to three different services, with 22.5 per cent providing four to six services, and 8.8 per cent providing seven to nine services. Ten percent of the sample did not provide any services at all, focusing instead on advice and signposting. Some

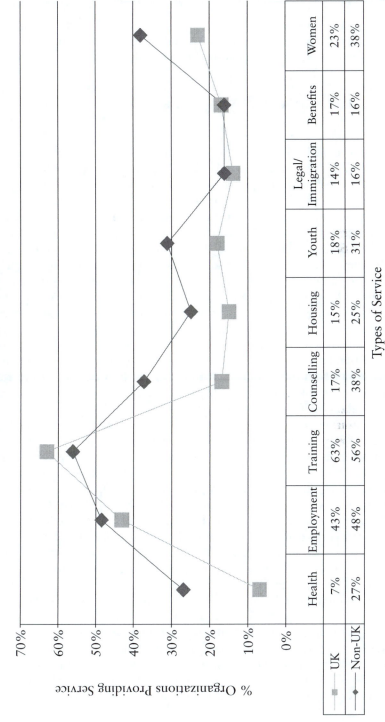

UK and Non-UK Service Provision

	Health	Employment	Training	Counselling	Housing	Youth	Legal/ Immigration	Benefits	Women
UK	7%	43%	63%	17%	15%	18%	14%	17%	23%
Non-UK	27%	48%	56%	38%	25%	31%	16%	16%	38%

Types of Service

% Organizations Providing Service

Figure 13.2 Comparison of UK and non-UK service provision.

58 per cent of organisations monitored the outcomes of all their efforts. Local organisations were the least likely to monitor whether their clients had entered employment or training as a result of their intervention. National integration programmes in Scandinavia were independently evaluated.

EMPLOYERS AND WORK EXPERIENCE

Providing links to employers was one of the key functions of the organisations surveyed, with 72.3 per cent of UK and 59.4 per cent overseas organisations fulfilling this role. Organisations operating in the sector over five years were most likely to offer support in this area. Interviews with organisations who arranged work-experience places revealed that in most cases the UK organisations operated on an ad hoc and very small-scale basis, perhaps responding to the request for work experience from the occasional refugee rather than actively running a structured work-experience programme. Whilst they recognised the importance of work experience in enhancing refugees' employability, they simply did not have the resources to be more proactive. Only four organisations were identified in the UK who operated structured work-experience programmes. This type of approach appeared to be more widely available in Europe. For example, in Scandinavian countries workless refugees were compelled to take up a range of work-experience places. In Holland, most colleges that provided language training for refugees also ran vocational courses that offered work-experience placements. Other organisations offered work placements relating to specific skills areas such as care, hospitality and health.

Organisations conceded that gaining commitment from employers to provide work placements for refugees was challenging. There were considerable differences in the levels of success experienced with some organisations, at an early stage in their activities, having had few successes, whilst others, such as one of the projects in Glasgow, now attracting more approaches from employers than they had capacity to deal with. Employers frequently confused the terms 'refugee' and 'asylum seeker', and much work was necessary to overcome some of the negative stereotypes that emerged from the popular press. Employers in the UK were also said to be extremely nervous of employment legislation, fearing that offering employment, even on a voluntary basis, might put them in breach of the law and attract a large fine (see Phillimore and Goodson, 2008). In the UK, work-experience programmes were set up in a fairly labour-intensive manner with designated project workers approaching employers to request a meeting to discuss placement provision. One organisation ran an initiative whereby they established work experience positions, wrote job descriptions and advertised, then interviewed for the posts. Once in post, trainees were offered a wide range of training to help boost their skills and provide them with enhanced job search knowledge. This approach appeared to have high success rates but operated at extremely low levels with an intake of no more than ten refugees per annum.

In Holland and Norway, organisations working with refugees were greatly assisted by the statutory nature of refugee integration programmes. For instance, in Norway there was a statutory expectation that local labour offices work with colleges running integration and language programmes to provide work placements. In Holland, the technical colleges who were contracted to provide integration programmes as part of state initiatives placed students on their standard, long established work-experience initiatives. In France, refugees often worked for organisations receiving support under the Insertion Companies initiative. Under this programme, they could work for a maximum of two years and receive the minimum wage whilst gaining real work experience. In Germany, although there were mechanisms for colleges to work with employers to set up work-experience places, employers were only allowed to employ a 'migrant' if they could demonstrate that they had been unable to recruit a German national. Only one organisation in the UK operated a work-placement programme that tried to address a skills gap. There was some consensus from UK organisations that, properly organised, such an approach could be less resource intensive than trying to match refugees with placements on a one-to-one basis. The orientation of programmes to skills gaps was much more common in Holland, Norway and Germany. Norwegian municipalities worked with the Labour Office and large local employers to build up a bank of placements from which refugees could choose, providing their skills were appropriate.

Support

High levels of support were found to be necessary for both employer and client. However, getting the balance right was considered 'tricky'. In Norway, the support provided was very intensive and the FAFO Integration Programme evaluation report concluded, 'There is a fine line between support and stalking' (Lund, 2006). They argued that whilst it was acceptable to probe refugees about their progress, the provision of a dedicated case worker achieved better results. The worker dealt with difficulties on an ad hoc basis, and this helped empower refugees to solve their own problems so that the level of support needed reduced over time. Providing the right support to employers was considered critical. Employers who did not have their questions answered and problems solved immediately quickly withdrew from initiatives. The research demonstrated that work experience required far more than simply placing a refugee in the workplace. Proper assessment and support were essential to ensure success. Operating successful work-experience initiatives demanded high levels of skills and knowledge. These resources were not available to many organisations and had often been purchased in the UK with the aid of specific funding from local authorities, the Scottish Executive, competitive funds (i.e. Single Regeneration Budget or Challenge funds) or the Home Office. Outside of the UK, core funding for these activities meant that a more strategic approach to placement provision and support could be established and sustained. The skilled nature of the support role was reflected in the decision in Norway to professionalise the integration officer role.

ACCREDITING PRIOR LEARNING AND EXPERIENCE

Research has indicated that a sizeable proportion of refugees arriving in the UK possess high levels of skills and education (e.g. Bloch, 2002; Aldridge et al., 2002; Phillimore and Goodson, 2006). For example, Kirk (2004) found that 23 per cent of refugees who provided details of their occupation before coming to the UK worked in skilled professions and a further 22 per cent were managers or senior officials. This compares to 12 per cent skilled and 15 per cent managers amongst UK-born people. The employability-initiatives questionnaire demonstrated quantitatively what a number of studies undertaken for the Learning and Skills Councils had revealed qualitatively, that there is much reliance on the UK NARIC system of qualification translation in the UK (see Phillimore and Goodson et al., 2003, 2004; Goodson and Phillimore, 2005). This is problematic for refugees because few have their paper qualifications with them, and the system tends to cover mainly academic qualifications and to downgrade other skill qualifications (ibid.; Mondi, 2005). Some 46 per cent of the UK organisations (56% if all countries were included) surveyed said they were involved in some form of work to help migrants to get the skills and/or qualifications that they had received in their home countries recognised in their new country. The forms of skill/qualification recognition that organisations said they were employing included qualification translation, qualification conversion, and accrediting or certifying skills/experience. However, interviews revealed that out of the sixteen organisations in the UK that reported they were using this technique, only three were actually accrediting prior learning, and only one of these initiatives was providing refugees with a certificate with an accrediting body; the others provided a certificate listing their skills. All others interviewed were using the UK NARIC system to translate academic qualifications.

In order to explore the ways in which refugees' skills can be accredited, we need to look to the example located in Scotland and the experience of overseas organisations. A refugee employability project based in Glasgow established an initiative to help construction workers requalify. This project combined vocational ESOL with workplace orientation; training to get students up to UK standard; and practical tests which enabled any construction worker to have their skills accredited in the National Vocational

Qualification system. Refugees were sent out to work-experience places to take their final practical tests and once successful could apply for their health and safety and union cards.

The Norwegian Ministry of Education and Research was asked by the Integration Department to explore the ways in which skilled and experienced migrants might be assessed in order to provide them with certificates accrediting their abilities. The Vocational Assessment Act of 2002 gives all new arrivals in Norway the opportunity to do an assessment of their nonformal and informal learning. They have developed and evaluated three techniques: vocational testing, a dialogue approach and assessment of portfolio. The Netherlands is also developing a portfolio approach for refugees, which will enable them to document all their different skills and qualifications in a very detailed manner. At the time of writing this system has not been linked with a formal credit system.

The system that enables refugees to gain a certificate from an examination board recognised by employers is the vocational-testing system. In Norway, vocational 'testing' starts off with an interview, where the background, training, work experience, language skills and objective of the adult are charted. After the first general interview a professional specialist interviews the individual in the particular subject, after which the individual demonstrates his or her abilities in practice, so that both the theoretical and the practical side of the trade is assessed. Working on the basis of this practice, the adult may be offered either additional education to bring them up to a trade certificate level or a public certificate useful for job seeking. This method complements other methods in that assessment of nonformal learning is also possible, and where required, parts or all of the practical side of the vocational subjects can be approved. Vocational 'testing' provides adults with an opportunity to show what they can actually do in their own fields. The method identifies knowledge and experiences that are not documented and works well, irrespective of learning and language difficulties. Both manual and computerised tools have been developed and tested in vocational and general subjects. The tools are used in different ways in the different methods, dependent on the needs of the individual. Sometimes the assessor supplements the existing tools with locally developed tools. A refugee's municipality covers the costs of the vocational tests. Some of the Dutch colleges also employ a similar system.

A further approach is commonly used in France, Germany and, on some courses with a work-experience dimension, in Holland. It involves the use of testing within the workplace, as refugees demonstrate their competence by actually undertaking a job. France has a particularly well-established system. This commences with a day evaluation. If the refugees pass the evaluation, then they can be sent to a company to undertake some unpaid work experience (maximum of 15 days) on which they can be tested. The National Employment Agency pays for an assessor, and where relevant an interpreter, to accompany the individuals. They undertake a range of practical, work-based exercises. If the skills are of the same or similar level to the French equivalent, then they are given a certificate. The certificate is known as an Evaluation Sheet and details their skills and equivalence. If their skills are deemed inadequate for the refugee to work in a particular field, they are recommended for training and must join general waiting lists. France has also developed software in a range of languages that can help skills to be identified prior to the work-based testing. There is a similar work-based process in Germany but it is lengthier. Refugees are observed at three three-weeklong placements. Where appropriate, certificates are issued at the end of the period.

There are a number of differences between the UK and the countries visited in the EU. In Holland, France and Germany there are long-established Accreditation of Prior Experiential Learning (APEL) systems that are open to the general population. In Scandinavia, systems are newer but have been extensively evaluated to determine the types of approaches that are effective and the target groups for each technique. In the UK, we have no national APEL system as such. There are practical tests for individuals working for registered construction employers and National Vocational Qualifications (NVQs) for those who have managed to find a job in an area where they have skills. There are no systems for unemployed individuals who wish to have their skills and abilities tested when they are outside of employment. Furthermore, considerable levels of resources are available in the EU for APEL. These are provided centrally as part of the funds earmarked for integration. Organisations in the UK must compete for local or national funds and thus have only been able to develop small-scale APEL programmes that work with small numbers of people in small geographical areas.

Other Approaches

Some 74 per cent of organisations surveyed were involved in other activities to try to enhance refugees' employability. Types of activity included liaising with employers, training schemes for employers, open days and joining networks to promote refugees. Interviews identified that the majority of organisations simply did not have the capacity to undertake much in the way of developmental or promotional work. One national-level organisation in the UK was working to develop employer toolkits to help demystify the employment legislation. Another UK organisation had set up an agency and some training to help refugees offer their services as translators and buddied novice translators with more experienced individuals. The Dutch Refugee Council and local municipalities came together in partnership to set up a mentoring organisation that trained retired business people to mentor refugees seeking work and to locate job opportunities for them. Some 50 per cent of refugees assisted were successful in finding work because of this intervention.

CONCLUSION

This research has explored the types of employability services available to refugees in the EU. The questionnaires were dispatched to organisations largely identified through Internet search, networking or snowballing and thus could be argued to be skewed towards larger organisations with sufficient resources to develop a Web site. However, the site visits confirmed the predominance of RCO provision in the UK and a more centralised approach elsewhere in the EU. Some key differences were identified between the approaches undertaken in the UK and Northern Europe. Many of the organisations trying to support refugees to enhance their employability in the UK were operating at a relatively small scale and trying to help refugees into work in an ad hoc fashion, developing relationships with employers on a needs basis rather than in a strategic manner. There were marked differences in the UK between the national-based organisations, which tend to be located in London, and those smaller, more recently established organisations that were found in the dispersal areas. However, wherever they were located and whatever their size, there were few programmes designed specifically to enhance refugees' employability. UK organisations reported a range of difficulties in trying to develop employability services. While they recognised the key importance of skills recognition and work experience, they lacked the capacity, space and expertise to deliver such services. Their efforts tended to be focused on dealing with refugees' immediate needs such as finding housing or getting help with a health problem.

Work-experience programmes for refugees are not new to the UK. However, the numbers of refugees offered placements has been very small compared with the number of unemployed refugees seeking work. Often, placement provision is within an NGO, and does not enable refugees to match their skills to those of an employer in order to increase the likelihood of return to their previous career. Furthermore, few organisations have the resources to offer both employer and employee the support they need to ensure that both get the most out of the experience. The situation is very different elsewhere in Northern Europe. Here there are well-established work experience programmes, which, with some extra support, refugees have been able to tap into, or newly developed programmes targeted specifically at refugees. This research has indicated that a number of strategic approaches have been developed in Northern Europe to help develop refugees' employability. Whilst in the UK little systematic monitoring takes place, elsewhere in the EU monitoring is more common and gives us some sense of what approaches are effective. In Scandinavia and the Netherlands, the main aim of integration programmes is to get refugees into sustainable employment. They use a combination of intensive language training, vocational orientation, training, accreditation of existing skills and experience and work experience coupled with intensive one-to-one support for refugees. Unlike the support available in the UK, which is resourced by short-term, often competitive funding, these EU programmes are core funded. Such initiatives are embedded in the work of statutory agencies and local municipalities, both of which have an obligation to deliver services and demonstrate success.

Rather than being supported on an ad hoc basis by overstretched, underfunded voluntary organisations, refugees in Northern Europe are supported within mainstream organisations by professionals. It is perhaps unsurprising that refugee unemployment rates are so high in the UK when we consider that little in the way of resources is invested in refugee employability initiatives. Some of the European initiatives have been very successful in helping refugees to gain work. Whilst they are resource intensive in the first instance, it is recognised that the savings made in benefits and support and the filling of vital skills gaps mean that in the medium to longer term the initiatives more than pay for themselves (Lund, 2006). However, while UK organisations were frustrated by the lack of funds to help them to develop employability services, larger organisations outside of the UK expressed concern at the lack of refugee-led organisations able to work with them to shape service provision. They viewed the UK's more bottom-up approach to service provision as a strength that was not being effectively utilised.

While some organisations in the UK at least attempted to locate work-experience places, there have been few developments in terms of APEL for refugees in the UK. Only one respondent was able to assist with the recognition of skills and experience. Some of the Northern European countries have made considerable investment into APEL for migrants and can offer refugees the opportunity to have their skills tested either in centres or in unpaid work placements. Some also use a portfolio system and are developing mechanisms to use the document for assessment and employment purposes. Clearly there is a need to learn from the experiences of work experience and APEL initiatives both in the UK and Europe. In order to enable us to make the most of refugees' skills and experience, the UK needs to develop a national system to recognise the work experience and skills of people from overseas. Creating an effective and meaningful system requires substantial levels of funding. All the methods documented earlier offer some potential to the UK. The dialogue and vocational-assessment systems are likely to be most difficult to implement because of the sheer volume of specialist assessors required and the lack of specialist centres. Perhaps the most practical approach is that of work-based skills assessment. This would give those refugees with largely practical skills a rapid solution to requalification and would not require the development of specialist test centres. It might be developed in association with work-experience initiatives to enable refugees to accredit their skills whilst gaining UK work experience, a reference and all the other benefits associated with work placements. It might make use of the existing assessors whose work is currently focused upon school leavers or individuals in skilled employment. It is unlikely that any progress will be made on the development of a national APEL system unless an organisation is charged with the responsibility for its development and allocated sufficient funds for the training and systems development that are necessary. The overall responsibility for such work might lie with the Learning and Skills Council, the organisation with responsibility for post-compulsory education. However, the development of such a system should be undertaken in association with refugee organisations and employer groups to ensure the mechanism developed is attractive to all parties. Systems can only be effective if they are recognised by employers. A starting point would be consultation with those employers who have a social-responsibility agenda to explore the ways in which work experience and skills recognition could be made attractive for employers.

Work experience and APEL are critical to enabling refugees to gain employment commensurate with their skills. Employment is central to integration. It is important in the long term that there is a commitment to funding work experience. Perhaps there might be a statutory obligation for organisations such as colleges and job centres to work together to provide work-experience activity following the example of the Scandinavian models. Introducing a statutory funding element to integration and work-experience and APEL activity would enable it to be provided wherever there were unemployed refugees. However, although the importance of employment to integration is recognised, it is only one aspect of integration. Commentators recognise that integration is made up of a multilayered set of processes often characterized by a combination of complex and interrelated issues (Castles et al., 2003; Fyvie et al., 2003).

Successful integration programmes need to take account of the wideranging nature of integration and address some of the other issues affecting an individual's ability to integrate, and often to work, including other functional aspects such as housing and access to services as well as issues around social capital. While organisations in Northern Europe were making progress addressing refugee unemployment, they recognised that greater success could be achieved through working with refugee communities to develop approaches that tackled all their integration needs.

The year 2008 will see the development of a national integration programme in the UK which, like those elsewhere in the EU, will place a great deal of focus upon employability. This development, although only aimed at new refugees, offers a significant opportunity to bring together the centralised approach developed in Northern Europe with the RCO-led approach currently employed in the UK. At the present time all indications are that the service will rely on signposting to state services and will be offered by private-sector advice and guidance suppliers rather than developing new services with the aid of RCOs. This chapter has demonstrated that work experience and APEL have key roles in helping to overcome some of the barriers that skilled refugees have in contributing to society to the extent of their potential. With joint working and sufficient funding it will be possible to bring structured, quality programmes to the UK. We can only hope that the new service does not miss the opportunity to build upon existing good practice.

NOTE

1. It was necessary to use the term 'migrants' in the EU because some countries did not recognise the term 'refugee' and combined all newcomers under the term 'migrant'.

14 The Future of Work for Recent Migrants and Refugees

Sonia McKay

The preceding chapters have vividly described the impact of globalisation, bringing with it the 'push' imperative of migration, from East to West and from South to North. The advanced industrialised countries of Europe, North America and Australasia have both 'welcomed' and deterred migrant workers, sending out mixed messages that are 'encouraging' of certain forms of migration, at the same time as they are strengthening national borders and tightening immigration controls. This chapter therefore provides an opportunity to review the findings and arguments raised in the preceding chapters and to move towards an overall analysis of why there are barriers to successful labour market integration and what measures it might be possible to adopt to address them. The chapter thus reflects on some of the key issues raised in the book. First, it reviews whom we are describing when we talk about migration, migrants and refugees. It then focuses on the impacts of racism and gender on the experiences of migration. It reviews the extent to which the theories of human and social capital provide an appropriate or adequate account of the migrant experience, particularly where it is immersed in low-skilled, low-paid, marginal labour. The chapter then turns to consider the role of the state, its responsibilities and the impact of its immigration policies on migrants. It concludes by reflecting on whether migration could represent a positive vehicle for change through its capacity to challenge discrimination and to forge effective engagements between host and new communities.

The different contributors to this book make it clear that while there is no such thing as a 'typical' migrant and that migrants are a heterogeneous group, they do share common experiences, although these can be affected by gender and ethnicity. The reasons as to why they have migrated and, in addition, the work that they carry out postmigration resonate from country to country and from one migrant group to another. Whether from Los Angeles to Sydney, or from London to Turin, the book chronicles similar stories; of people forced to migrate, whether for economic or political reasons (and more frequently through a combination of both); of individuals with higher levels of skills and qualifications than the average for the countries they migrate to; and yet almost universally of individuals occupying the same lower skilled, lower paid jobs than their qualifications would indicate, whether in the fields of California or in the food-processing factories of the east of England. The question is not why their demographic makeup is so similar, since it might be expected that the same characteristics that make individuals migrate to the USA would also be present in those who migrate to France. However, what is of note is that the predicaments they face postmigration are so similar. Thus, while it might be assumed that states with differing political heritages and politics would offer a more varied range of possibilities for recent migrants and refugees, the book shows that this is not the case. The fact that they do not, that, for example, the experiences of refugees are similar whether they have fled to the UK, to Australia or to Italy, suggests that there are wider interests that determine the conditions under which they are received in destination countries. Giovanni Mottura and Matteo Rinaldini, in Chapter 6, discuss the Marxist theory of the reserve army of labour, in relation to migration to Italy, but it is easy to view the theory as having equal application to the predicament of migrants and refugees in all of the countries covered in this book.

The book's contributors have often drawn analogies between the positions of migrants and refugees, arguing that there are more issues that unite them than that separate them. While none suggests that the two statuses are completely interchangeable, nevertheless they tend to a conclusion that in many cases an individual's migration status is determined by the policies of the receiving state as much as by the migrant's objective situation.

The book has focused primarily on the experiences of those who have recently migrated, although Jefferys, in his chapter on France, reflects on migration in both the pre- and postcolonial periods. It also needs to be stated that, while an individual's status may be allocated at the point of time of arrival, how migrants and refugees are perceived and also how they perceive themselves are dependent on the extent to which they find acceptance within the host society. And acceptance is often predicated on notions of familiarity, meaning that host societies, such as those described in this book, may be more ready to view white migrants and refugees as 'acceptable', thus rejecting for them the labels of 'migrant', 'refugee' and 'immigrant'. For visibly different migrants, however, these labels are more difficult to cast off. Indeed, as Jefferys and other contributors have noted, the definition of an 'immigrant' can be attached to second and future generations, in the case of those perceived as visibly different. Val Colic-Peisker similarly shows that visible difference is a key element in any attempt to understand why it is that some groups of migrants and refugees have particularly poor labour market outcomes.

RACISM AND SEXISM AS KEY DETERMINANTS

For these reasons it is not possible to discuss contemporary migration without considering the impacts of racism and sexism, which determine, to a greater or lesser extent, where individuals will be located within the labour market. As the contributors to this book have eloquently demonstrated, racism encourages labour market segregation and limits the powers of individuals to challenge discriminatory and racist practices where immigration status or limited job opportunities place them in positions of vulnerability. Additionally, in a post-9/11 world, we know that migrants who are visibly different experience greater levels of hostility, both in workplaces and in wider communities. This forces individuals back into reliance on their own immediate support mechanisms, and while it may allow individuals to 'survive' in their new locations, it rarely provides access to high-quality, decent jobs.

Female migrants and refugees may not only experience racism in host countries but are also trapped by gender stereotypes that limit them to particular sectors of employment—in the care industry, in hotels and catering and in cleaning—and the fact that this pattern seems to repeat itself with depressing similarity in all of the countries covered in this book is of note. As Val Preston and Silvia D'Addario, in their chapter on migration into Canada show, sexism and racism make the labour market experiences of women migrants even more challenging. Migration for women therefore is not a liberating experience, in the sense that it allows them to escape the narrow stereotyping of what is seen as women's work. Instead, it reinforces occupational segregation and pigeonholes women migrants even more resolutely into women's work. However, for some, it can at the same time also represent a route to a 'valued' economic status, as primary household earners, as, for example, in the case of nurses from the Philippines working in the UK healthcare sector. Women may also be more 'conditioned' to undertake compromises to survive and it may also be that the opportunity to engage in paid work outside the home is itself viewed as a positive outcome of migration. As a consequence, women, when asked to give an account of their migration experience, may offer a more positive assessment of their postmigration predicament than do male recent migrants, as migration may represent either a curtailment or a liberation, dependent on what parts of the labour market she is located in postmigration and the extent to which she is able to make choices.

SOCIAL AND HUMAN CAPITAL

Val Colic-Peisker, in reviewing the employment experiences of refugees into Australia, notes that they use their social capital to access employment. This

results in jobs mainly in ethnic firms that provide little employment opportunity for the highly skilled. Furthermore, she finds that ethnic entrepreneurs often consciously employ newly arrived coethnics, expecting them to
be cheap and pliable labour. She refers to similar findings from Canada
(Lamba, 2004). Giovanni Mottura and Matteo Rinaldini make a similar
assessment. Migrants into Italy, utilising their social capital, find that jobs
are accessed through networks of existing migrant employees, which is not
only a 'low-cost' option for employers but results in 'both an inclusion and
exclusion process' which locates specific nationalities within limited job
enclaves. From these and other contributions, it is possible to state that
social capital does not automatically provide access to decent work and that
social networks, while representing channels into work, at the same time are
traps in work.

The theoretical model of human capital assumes that the skills and qualifications that individuals have place them in an advantaged position in the
labour market. Applying this theory to refugees and migrants should suggest
that since, by and large, they 'possess' higher human capital, they should be
advantaged in the labour markets of their destination countries. However,
again, as this book's contributors have shown, there is no necessary direct
relationship between the human capital that migrants bring and the jobs
that they find themselves doing. Nor is it the case that it is merely a question of individuals' not having 'recognised' human capital. Allan Williams
provides a forceful argument, in his chapter on employability, that even taking account of the fact that migrants may enter the host labour market
suboptimally, because they lack the necessary local human capital, 'it does
not automatically follow that the acquisition of nationally specific human
capital over time will lead to higher wages and occupational mobility for
the migrants'. Williams argues that whether or not this occurs will depend
on whether 'these initial jobs constitute labour market stepping-stones or
labour market entrapments' and, as contributors have shown, labour market entrapment is a more common experience, particularly for those who
are visibly different. The lack of relationship between human-capital accumulation and job success is also a point made in my chapter on access to
work, which notes that labour market success, if measured by the ability to
obtain work, is actually better for those with relatively low human capital,
while for those with higher human capital there may be greater barriers
to accessing work. Colic-Peisker's research purposely focused on refugees
with relatively high human capital, yet she too found that employment outcomes for this group after five to seven years of residence were poor. For
her, the conclusions were that 'human capital itself cannot compensate for
labour market segmentation, which allocates the visibly different refugees
into undesirable jobs' and that labour market segmentation resulting from
systemic discrimination and reinforced by mainstream prejudices and negative stereotyping of racially and culturally different immigrants and refugees by employers is the principal reason for poor labour market outcomes.

Alice Bloch similarly, in her work on refugees, found a limited correlation between high human capital and successful labour market outcomes. Again, race and gender were key determinants.

STATE POLICIES AND STATUS ISSUES

A common issue explored in each of the chapters in this book is how state policies have inexorably moved towards tightened migration control and towards the 'management' of migration, through the imposition of quotas and other restrictions on entry. While migration is acknowledged as a necessary response to labour and skills shortages and to an increasingly aging population in most Western economies, government policies have pushed greater numbers of workers into noncompliant employment. Migration in the early twenty-first century is increasingly identified with irregular employment, as a result of insecure or illegal migration status. And this is not accidental. Governments argue that they need stricter controls to restrict or regulate entry, but, as Nandita Sharma points out, 'restrictive migration policies have not worked to restrict people's mobility'. What they have done is to drive that labour into the most exploitative employment relationships. Thus, we need to question why it is that state policies have proven to be so spectacularly unsuccessful, so that even without precise figures, millions of workers are estimated as being present in unregulated employment, mainly because they have no authorisation either to work or to reside in the host country.

Most of the book's contributors thus conclude that state policies are themselves a key determinant of poor labour market outcomes, and furthermore this is not accidental but purposeful. Sharma, in her chapter on the USA, cites Rivera-Batiz (1999), who found that the main reason for unequal wages between migrants and indigenous workers was related to their irregular status and that human capital differences had marginal affects status. Anderson (2007) shows that there is indeed a dynamic interrelationship between immigration controls and precarious labour. This supports the conclusion that the creation of a large and growing global undocumented labour force is not accidental but has been constructed. Preston and D'Addario demonstrate that state policies focusing on skilled migrants do not deliver jobs in skilled areas for migrant women.

Formal state policies in all of the countries examined are said to support integration models and to enhance employability. But in practice there is limited evidence that such policies have had sufficient priority and resources to make them viable. As Jenny Phillimore demonstrates in her research on cross-European models of integration, the enhancement of employability and support for integration is handled piecemeal in many European countries, with limited funding and the lack of coherent strategies to address the multiple issues that migrants and refugees encounter, including access to training, accommodation and health care, as well as to employment.

MIGRATION AS A FOCUS OF CHANGE

Migration has the potential to operate as an important agent of change. Migrants and refugees bring with them new ideas and experiences and have throughout history been prime actors in major social movements for change. As Sharma records, the USA demonstrations on 1 May 2006 showed without doubt that not only can migrant workers be mobilised for change but also that they have the potential power to create or stimulate change. While this may not be in doubt, whether there exist at present, within host societies, social agents willing to engage with migrant workers is more open to question. In particular, for trade-union organisations in developed industrial economies, migration becomes a challenge. It is only if they and other social actors are capable of utilising the enthusiasm and energy of migrant labour that this could provide the key to rebuilding key organisation, not just among migrant workers but also among indigenous workers, who all too often are not themselves viewed as potential actors for social change.

Selected Bibliography

Adhikari, P. (1999). 'Are there migrant enclaves in Australia?—A search for the evidence'. *Australian Journal of Social Issues*, 14(3): 191–210.

Age, The. (2007/October 20). 'An indomitable spirit'. *The Age* (Melbourne), by Larissa Dubecki, *Insight*, p. 10.

Ager, A. & Strang A. (2004). *Indicators of integration: Final report*. London: Home Office, http://www.homeoffice.gov.uk/rds/pdfs04/dpr28.pdf.

Alboim, N., Finnie, R., & Meng, R. (2005). 'The discounting of migrants' skills in Canada'. *Institute for Research on Public Policy, Choices*, 11(2), 2–23.

Aldridge, F., Dutton, Y., Gray, R., et al. (2005). *Working to Rebuild Careers: An assessment of the provision to assist refugees seeking employment in the East Midlands*. Research Paper Employability Forum: London.

Aldridge, F., & Waddington, S. (2001), *Asylum seekers' skills and qualifications audit pilot project*. Leicester, UK: National Organisation for Adult Learning.

Ambrosini, M. (2001). *La fatica di integrarsi: Immigrazione e lavoro in Italia*. Bologna: Il Mulino.

Amin, A. (1994). *Post-Fordism: A reader*. Oxford: Blackwell.

Amin, A. (2002). 'Spatialities of globalization'. *Environment and Planning A*, 34: 385–99.

Anderson, B. (2007). *Battles in time: The relationship between global and labour mobilities*, Working Paper No. 55, Compas. Oxford: University of Oxford.

Anderson, B., Ruhs, M., Rogaly, B., et al. (2006). *Fair enough? Central and East European migrants in low-wage employment in the UK*. York, UK: Joseph Rowntree Foundation.

Andolfatto, D. (2007). Introduction, *Les syndicats en France*. Paris: La Documentation Française, 9–14.

Anker, D. (2002). 'Refugee law, gender and the human rights paradigm'. *Harvard Human Rights Journal*, 15: 133.

Arber, S. (1993). 'The research process'. In Gilbert. G. N. (ed.), *Researching social life*, pp. 32–50. London: Sage,

Archer, L., Hollingworth, S., Maylor, U., et al. (2005). Challenging barriers to employment for refugees and asylum seekers in London. London: Institute for Policy Studies in Education, London Metropolitan University.

Armstrong, P., & Armstrong, H. (2001). *The double ghetto: Canadian women and their segregated work* (3rd ed). Toronto: Oxford University Press.

Arthur, M. B., & Rousseau, D. M. (1996). *The boundaryless career: A new employment principle for a new organizational era*. New York: Oxford University Press.

Asylum Coalition, The. (2002/September). *Asylum City*. http://www.npi.org.uk/reports/asylum%20city.pdf

240   *Selected Bibliography*

Atkinson, J. (1984). *Flexibility, uncertainty and manpower management*, Report 89. Brighton, UK: Institute of Manpower Studies, University of Sussex.

Audit Commission. (2007). *Crossing borders: Responding to the local challenges of migrant workers*. London: Audit Commission.

Australian Productivity Commission. (2002). *Independent review of the Job Network*, Report No. 21. Canberra: AusInfo.

Aydemir, A. (2002). 'Effects of selection criteria and economic opportunities on the characteristics of migrants. *Analytical studies research paper no. 192*. Ottawa: Statistics Canada.

Aydemir, A. (2003). 'Effects of business cycles on the labour market participation and employment rates assimilation of migrants'. In C. M. Beach, A. G. Green, & J. G. Reitz (eds.), *Canadian immigration policy for the 21st century*, pp. 373–412. Kingston: John Deutsch Institute for the Study of Economic Policy, Queen's University.

Aydemir, A., & Skuterud, M. (2004). 'Explaining the deteriorating entry earnings of Canada's migrant cohort: 1996–2000'. *Analytical studies research paper*. Ottawa: Statistics Canada.

Bacon, D. (1999). 'Immigrant workers ask labor—"Which side are you on?"' posted Novemeber 18, 1999, http://dbacon.igc.org/Imgrants/23WhichSide.htm, accessed December 22, 2007.

Badets, J., & Howatson-Leo, L. (1999). 'Recent migrants in the workforce. *Canadian Social Trends*, Spring, 16–23.

Bakan, Abigail B., & Stasiulis, D. (1997). *Not one of the family: Foreign domestic workers in Canada*. Toronto: University of Toronto Press.

Baldaccini, A. (2003). *EU and US approaches to the management of immigration: United Kingdom* [online]. Jan Niessen, Yongmi Schibel and Raphaele Magoni (eds.). Brussels: Migration Policy Group, www.migpolgroup.com.

Baldwin, J. R., & Beckstead, D. (2003). *Knowledge workers in Canada's economy, 1971–2001*. Ottawa: Statistics Canada.

Barley, S., & Kunda, G. (2004). *Gurus, hired guns, and warm bodies: Itinerant experts in a knowledge economy*. Princeton, NJ: Princeton University Press.

Barth, F. (1994). 'Enduring and emerging issues in the analysis of ethnicity'. In *The anthropology of ethnicity: Beyond 'ethnic groups and boundaries'*, H. Vermeulen & C. Govers (eds.), pp. 11–31 Amsterdam: Het Spinhuis.

Bassnet, S. (1994). *Translation studies*. London: Routledge.

Basso, P., & Perocco, F. (2003). 'Gli immigrati in Europa'. In P. Basso & F. Perocco (eds.), *Gli immigrati in Europa: Disuguaglianze, razzismo, lotte*, pp. 7–54, Milano: FrancoAngeli.

Bataille, P. (1997). *Le racisme au travail*. Paris: La Découverte.

Bate, R., Best, R., & Holmans, A. (2005). *On the move: The housing consequences of migration*. York, UK: Joseph Rowntree Foundation

Baubock, R. (1991). *Immigration and the boundaries of citizenship*. Warwick, UK: Centre for Research in Ethnic Relations.

Baum, T., Dutton, E., Karimi, S., et al. (2007). 'Cultural diversity in hospital work'. *Cross Cultural Management*, 14(3): 229–39.

Beck, U., & Beck-Gernsheim, E. (2002). *Individualization*. London: Sage.

Becker, G. (1975). *Human Capital* (2nd ed.). New York: Columbia University Press.

Begum, N. (2004, June). 'Employment by occupation and industry'. *Labour Market Trends*, pp. 227–34, http://www.statistics.gov.uk/cci/article.asp?id=905.

Behdad, A. (2005). *A forgetful nation: On immigration and cultural identity in the United States*. Durham, NC, & London: Duke University Press.

Beiser, M., & Hou, F. (2000). 'Gender differences in language acquisition and employment consequences among Southeast Asian refugees in Canada'. *Canadian Public Policy*, 26, 3, 311–30.

Bentley, T. (1998). *Learning beyond the classroom: Education for a changing world*. London: Routledge.

Bertaux, D., & Kohli, M. (1984). 'The life story approach: A continental view'. *Annual Review of Sociology*, 10: 215–37.

Berthoud, R. (2000). 'Ethnic employment penalties in Britain'. *Journal of Ethnic and Migration Studies*, 26(3): 389–416.

Birch, M., & Miller, T. (2002). 'Encouraging participation: Ethics and responsibilities. In M. Mauthner, M. Birch, J., Jessop, et al. (eds.), *Ethics in qualitative research*, pp. 91–106. London, Sage.

Bloch, A. (1999). 'Refugees in the job market: A case of unused skills in the British Economy'. In A. Bloch & C. Levy (eds.), *Refugees, citizenship and policy in Europe*, pp. 187–210. Basingstoke, UK: Palgrave.

Bloch, A. (2000). 'Refugee settlement in Britain: The impact of policy on participation'. *Journal of Ethnic and Migration Studies*, 26(1): 75–88.

Bloch, A. (2002). *Refugees' opportunities and barriers in employment and training*, DWP Research Report 179. Leeds, UK: CDS.

Bloch, A. (2002). *The migration and settlement of refugees in Britain*. Basingstoke, UK, & New York: Palgrave Macmillan.

Bloch, A. (2004a). 'Survey research with refugees: A methodological perspective'. *Policy Studies*, 25: 2, 139–51.

Bloch, A. (2004b). *Making it work: Refugee employment in the UK*. London: Institute of Public Policy Research.

Bloch, A. (2007). 'Methodological challenges for national and multi-sited comparative survey research'. *Journal of Refugee Studies*. 20(2) 230–247.

Bloch, A. (2008). 'Refugees in the UK labour market: The conflict between economic integration and policy-led labour market restriction'. *Journal of Social Policy*, 37(1) 21–36.

Bohning, W. R. (1967). *International labour migration*. London: Macmillan.

Bohning, W. R. (1972). *The migration of workers in the UK and the European Community*. Oxford: Oxford University Press.

Bonacich, E. (1979). 'The past, present and future of split labor market theory'. In C. B. Marrett & C. Leggon (eds.), *Research in race and ethnic relations* (A research annual). Greenwich, CT: JAI Press Inc.

Bonacich, E., & Modell, J. (1980). *The economic basis of ethnic solidarity: Small Business in the Japanese American community*. Berkeley: University of California Press.

Borrel, C. (2006). 'Enquêtes annuelles de recensement 2004 et 2005'. *INSEE Premiere* (1098): 1–4.

Bowser, R. (2007). *Retention of ESOL students working in the food industry in South Lincolnshire*. University of Nottingham, Nottingham.

Boyd, M. (1992). 'Gender, visible minority and migrant earnings inequality: Reassessing an employment equity premise'. In V. Satzewich (ed.), *Deconstructing a nation: Immigration, multiculturalism and racism in 1990s Canada*, pp. 279–321. Halifax, Nova Scotia: Fernwood Publishing and University of Saskatchewan, Department of Sociology, Social Science Research Unit.

Brandenburg, D. C., & Ellinger, A. D. (2003). 'The future: Just-in-time learning expectations and potential implications for human resource development'. *Advances in Developing Human Resources*, 5(3): 308–20.

Breckner, R. (1998). 'The biographical-interpretative method—principles and procedures'. In *SOSTRIS Working Paper No. 2*, pp. 91–105. London: University of East London.

Brettell, C. B., & Hollifield, J. F. (2000). *Migration theory: Talking across disciplines*. New York and London: Routledge.

Brouwer, A. (1999). *Migrants need not apply*. Ottawa: Caledon Institute of Social Policy.

Canada. (2001). Immigration and Refugee Protection Act.

Brubaker, W. R. (1992). *Citizenship and nationhood in France and Germany*. Cambridge, MA: Harvard University Press.

Brusco, S. (1989). *Piccole imprese e distretti industriali*. Torino, Italy: Rosenberg & Sellier.

Bustamante, J., Clark, A., Reynolds, W., et al. (1992). *U.S.-Mexico relations: Labor market interdependence*. Stanford, CA: Stanford University Press.

Carchedi, F., Mottura, G., & Pugliese, E. (eds.) (2003). *Il lavoro servile e le nuove schiavitù*. Milano: FrancoAngeli.

Carey-Wood, J., et al. (1995). *The settlement of refugees in Britain*, Home Office Research Study 141 HMSO: London.

Caritas/Migrantes. (2005). *Immigrazione. Dossier statistico 2005. XV rapporto*. Roma: Anterem.

Caritas/Migrantes. (2007). *Immigrazione. Dossier statistico 2007. XVII rapporto*. Roma: Anterem.

Castles, S. (2007). 'Twenty-first century migration as a challenge to sociology'. *Journal of Ethnic and Migration Studies*, 33(3): 351–71.

Castles, S., Booth, H., et al. (1987). *Here for good: Western Europe's new ethnic minorities*. London, Pluto Press.

Castles, S., Korac, M., Vasta, M., et al. (2003). *Integration mapping the field*. Oxford: Oxford University Press.

CCVT (Canadian Centre for Victims of Torture). (2000). '*Befriending survivors of torture—building a web of community support*, www.ccvt.org/brifriendingwork shop.html.

Cediey, E., & Foroni, F. (2007). *Les discriminations à raison de « l'origine » dans les embauches en France: Une enquête nationale par tests de discrimination selon la méthode du Bureau International du Travail*. Geneva: Bureau International du Travail, 113.

Chambon, A., Abai, M., Dremetsikas, T., et al. (1998). *Methodology in university-community research partnerships: The Link-By-Link project, a case study*, www.ccvt.org/research.html.

Chard, J., Badets, J., & Howatson-Leo, L. (2000). 'Migrant women'. In *Women in Canada, 2000: A gender-based statistical report*, pp. 189–218. Ottawa: Statistics Canada.

Charliff, L., Ibrani, K., Lowe, M., et al. (2004). *Refugees and asylum seekers in Scotland: A skills and aspirations audit*. Edinburgh: Scottish Executive and Scottish Refugee Council.

Chiswick, B. R. (1978). 'The effect of Americanization on the earnings of foreign-born men'. *Journal of Political Economy*, 86: 897–921.

Chiswick, Barry R., & Miller, Paul W. (2002). 'Migrant earnings: Language skills, linguistic concentrations, and the business cycle'. *Journal of Population Economics*, 15, 31–57.

Chiu, T., & Zietsma, D. (2003). 'Earnings of migrants in the 1990s'. *Canadian Social Trends*, Autumn, 24–28.

Citizenship and Immigration Canada. (2003). *Facts and figures, 2002*. Ottawa: Citizenship and Immigration Canada.

Citizenship and Immigration Canada. (2006). *Facts and figures, 2005*. Ottawa: Citizenship and Immigration Canada.

Clifford, J. (1997). *Routes, travel and translation in the late twentieth century*. Cambridge, MA: Harvard University Press.

Colic-Peisker, V., & Tilbury, F. (2003). '"Active" and "passive" resettlement: The influence of support services and refugees' own resources on resettlement styles'. *International Migration*, 41: 61–91.

Colic-Peisker, V., & Tilbury, F. (2006). 'Employment niches for recent refugees: Segmented labour market in 21st-century Australia'. *Journal of Refugee Studies*, 19: 203–29.

Colic-Peisker, V., & Walker, I. (2003). 'Human capital, acculturation and social identity: Bosnian refugees in Australia'. *Journal of Community and Applied Social Psychology*, 13 :337–60.

Collins, J. (1991). *Migrant hands in a distant land* (2nd ed.). Sydney: Pluto Press Australia.

Commission for Rural Communities. (2007). *A8 migrant workers in rural areas—briefing paper*. Cheltenham, UK, & London: Commission for Rural Communities.

Commission on Integration and Cohesion. (2007). *Our shared future*. Wetherby, UK: Communities and Local Government Publications.

Commission on Integration and Cohesion. (2007a). *Our shared future*. Commission on Integration and Cohesion, http://www.integrationandcohesion.org.uk/upload/assets/www.integrationandcohesion.org.uk/our_shared_future.pdf.

Commission on Integration and Cohesion. (2007b). *Themes, messages and challenge: A summary of key themes from the Commission for Cohesion and Integration Consultation*. Commission on Integration and Cohesion, http://www.integrationandcohesion.org.uk/upload/assets/www.integrationandcohesion.org.uk/themes,_messages_and_challenges.pdf.

Commonwealth of Australia. (2007). *A comparison of Australian and Canadian immigration policies and labour market outcomes* (DIAC, October 2004), at http://www.immi.gov.au/media/publications/pdf/comparison_immigration_policies.pdf.

Connor, H., Tyers, C., Madood, T., et al. (2004). Why the difference? A closer look at higher education minority ethnic students and graduates, DfES Research Report RR 552, London.

Constable, J., Wagner, R., Childs, M., et al. (2004). *Doctors become taxidrivers: Recognising skill—not easy as it sounds*. Sydney: Office of Employment Equity and Diversity, Premier's Department, at www.eeo.nsw.gov.au.

Courtois, G., & Jaffré, J. (2001). 'La popularité des mouvement sociaux ne se dément pas depuis 1995'. *Le Monde*, p. 18 07 March 1995.

Coussey, M. (2000). *Framework of integration policies*, Director general III-Social Cohesion. Strasbourg: Council of Europe Publishing.

Cowles, K. (1988). 'Issues in qualitative research on sensitive topics'. *Western Journal of Nursing Research*, 10: 163–79.

Cox, R., & Watts, P. (2002). 'Globalisation, polarisation and the informal sector: The case of paid domestic workers in the UK'. *Area*, 31(1) 39–47.

Cranford, C., & Vosko, L. (2006). 'Conceptualizing precarious employment: Mapping wage work across social location and occupational context'. In L.Vosko (ed.), *Precarious employment*, pp. 43–67. Montreal: McGill-Queen's University Press.

Crisp, J. (1999). 'Policy challenges of the new diasporas: Migrant networks and their impact on asylum flows and regimes'. *Journal of Humanitarian Assistance*, Policy Research Unit UNHCR, Geneva, May 1999.

Cross, G. (1983). Immigrant workers in industrial France: The making of a new laboring class. Philadelphia: Temple University Press.

Daneo, C. (1971). *Agricoltura e sviluppo capitalistico in Italia*. Torino: Einaudi.

Dell'Olio, F. (2004). 'Immigration and immigrant policy in Italy and the UK: Is housing policy a barrier to a common approach towards immigration in the EU?' *Journal of Ethnic and Migration Studies*, 30(1): 107–28.

Delphy, C. (1984). *A materialist analysis of women's oppression*. Amherst, MA: University of Massachusetts.

Dench, S., Hurstfield, J., Hill, D., et al. (2006). 'Employers' use of migrant labour'. *Home Office Online Report* 03/06.

Department for Work and Pensions. (2005). *Working to rebuild lives: A refugee employment strategy*. Sheffield, UK: DWP.

Department for Work and Pensions and Department for Innovation, Universities and Skills. (2007). *Opportunity, employment and progression: Making skills work*, Cm 7288, Norwich, UK: The Stationery Office.

Department of Immigration and Multicultural and Indigenous Affairs. (2005). *Community information summary*, at www.immi.gov.

Department of Immigration and Multicultural and Indigenous Affairs. (2005a). *2006–07 humanitarian program—discussion paper*, at http://www.minister.immi.gov.au/consultations.

DFES. (2005). Skills in England—Volume 1. Coventry: Learning and Skills Council.

Diller, J. (1998). *In search of asylum: Vietnamese Boat People in Hong Kong*. Washington, DC: Indochina Resource Center.

Donolo, C. (1972). 'Sviluppo ineguale e disgregazione sociale nel Mezzogiorno'. *Quaderni Piacentini*, n. 47.

Dowling, S., Moreton, K., & Wright, L. (2007). 'Trafficking for the purposes of labour exploitation: A literature review'. *Home Office Online Report* 10/07.

Dudhwar, A. (2004). 'Towards a refugee employment strategy'. *Industrial Law Journal* 33(3): 286–90.

Dumper, H. (2002). *Missed opportunities: A skills audit of refugee women in London from teaching, nursing and medical professions*. London: GLA.

Dunn, L., & Somerville, W. (2004). *Barriers to employment*. London: Centre for Economic and Social Inclusion.

Dustmann, C., & Weiss, Y. (2007). 'Return migration: Theory and empirical evidence from the UK'. *British Journal of Industrial Relations*, 45(2): 236–56.

Edwards, R., & Ribbens, J. (1998). 'Living on the edges: Public knowledge, private lives, personal experience'. In J. Ribbens & R. Edwards (eds.), *Feminist dilemmas in qualitative research*, pp. 1–24. London, Sage.

Ehrenberg, R. G., & Smith, R. S. (1994). *Modern labor economics: Theory and public policy*. New York: HarperCollins.

Ehrenreich, B., & Russel, A. (eds.) (2004). *Donne globali: Tate colf e badanti*. Milano: Feltrinelli.

Einaudi, L. (2007). *Le politiche dell'immigrazione in Italia dall'unità ad oggi*. Bari, Italy: Laterza.

Employability Forum. (2006). *Rebuilding lives—groundwork progress report on refugee employment*. London: National Refugee Integration Forum Employment and Training Subgroup and Home Office.

English-Lueck, J. A., Darrah, C. N., & Saveri, A. (2002). 'Trusting strangers: Work relationships in four high-tech communities'. *Information, Communication and Society*, 5(1): 90–108.

Enneli, P., Modood, T., & Bradley, H. (2005). *Young Turks and Kurds: A set of 'invisible' disadvantaged groups*. York, UK: Joseph Rowntree Foundation.

Ensor, J., & Shah, A. (2005). 'United Kingdom'. In *Current immigration debates in Europe*, Niessen et al. (eds.). Brussels: Migration Policy Group.

ESOPE Project. (2004). *Managing labour market related risks in Europe: Policy implications* (final version). Financed by the European Commission, DG Research, V Framework Programme.

Esping-Anderson, G. (ed.) (1996). *Welfare states in transition: National adaptations in global economies*. London: Sage.

Esping-Anderson, G., & Regini, M. (eds.). *Why deregulate markets?* Oxford: Oxford University Press.

European Commission. (2001). *Report of the Third European Conference on the Integration of Refugees*. Brussels: European Commission.

Evans, M. D. R., & Kelly, K. (1991). 'Prejudice, discrimination, and the labor market: Attainments of immigrants in Australia'. *American Journal of Sociology*, 97(3): 721–59.

Farer, T. (1995). 'How the international system copes with involuntary migration: Norms, institutions and state practice'. *Human Rights Quarterly*, 17: 72.

Favell, A., & Hansen, R. (2002). 'Markets against politics: Migration, EU enlargement and the idea of Europe'. *Journal of Ethnic and Migration Studies*, 28(4): 581–602.

Feeney, A. (2000). 'Refugee employment'. In *Local Economy*, 15(4), 343–49.

Fischer-Rosenthal, W. (1995). 'The problem with identity: Biography as a solution to some (post)-modernist dilemmas'. *Comenius*, 15: 250–65.

Flynn, D. (2005). 'New Borders, new management: The dilemmas of modern immigration policies'. *Ethnic and Racial Studies*, 28(3), 463–90.

Foster, M. (2007). *International refugee law and socio-economic rights*. Cambridge: Cambridge Studies in International and Comparative Law.

Frenette, M., & Morisette, R. (2003). 'Will they ever converge? Earnings of migrant and Canadian-born workers over the last two decades'. *Analytical studies research paper*. Ottawa: Statistics Canada.

Fugazza, M. (2003). 'Racial discrimination: Theories, facts and policy'. *International Labor Review*, 142(4): 507–42.

Fyvie, A., Ager, A., Curley, G., et al. (2003). *Integration mapping the field, Volume II: Distilling policy lessons from the 'mapping the field' exercise*. Home Office Online Report 29/03.

Galarneau, D., & Morissette, R. (2004). 'Migrants: Settling for less?' *Perspectives on Labour and Income*, 5, 5–16.

Gallino, L. (2004). 'Globalizzazione della precarietà'. In I. Masulli (ed.), *Precarietà del lavoro e società precaria nell'Europa contemporanea*, pp. 9–25. Roma: Carocci.

Geddes, A. (2000). *Immigration and European integration: Towards fortress Europe?* Manchester, UK: Manchester University Press.

Gergen, M. (2001). Feminist reconstructions in psychology: Narrative, gender and performance. Thousand Oaks, CA: Sage.

Ghassan, H. (2000). *White nation: Fantasies of white supremacy in a multicultural society*. New York & Annandale, NSW, Australia: Routledge and Pluto Press.

Ghosh, B. (1998). *Huddled masses and uncertain shores: Insights into irregular migration*. The Hague: Martinus Nijhoff Publishers.

Gibbons, M. (2007). *A review of employment dispute resolution procedures in Great Britain*. DTI London.

Gibney, A. (1988). '"Well-founded fear" of persecution'. *Human Rights Quarterly*, 10(1): 109–21.

Gilbert, A., & Koser, K. (2006). 'Coming to the UK: What do asylum-seekers know about the UK before arrival?' *Journal of Ethnic and Migration Studies*, 32(7), 1209–25.

Giles, W., & Preston, V. (1996). 'The domestication of women's work: A comparison of Chinese and Portuguese migrant women homeworkers'. *Studies in Political Economy*, 51, 147–82.

Glover, S., Gott, C., Loisillon, A., et al. (2001). *Migration: An economic and social analysis*. RDS Occasional Paper No. 67 Home Office, London.

Gold, M., & Fraser, J. (2002). 'Managing self-management: Successful transitions to portfolio careers'. *Work, Employment and Society*, 16(4): 579–97.

Goldberg, D. (1993). Racist culture: Philosophy and the politics of meaning. Oxford: Blackwell.

Goodson, L., & Phillimore, J. (2005). *New migrants communities: Education, training, employment and integration matters*. Sandwell, UK: Learning and Skills Council Black Country.

Goodwin-Gill, G. (1996). *The Refugee in international law*. Oxford: Clarendon Press.

Goss, J., & Lindquist, B. (1995). 'Conceptualizing international labour migration: A structuration perspective'. *International Migration Review*, 29(2): 317–51.

Gouldner, A. (1971). *The coming crisis of Western sociology*. London: Heinemann.

Granovetter, M. (1973). 'The strength of weak ties'. *American Journal of Sociology*, 78: 1360–80.

Green, A. (2006). 'Routes into employment for refugees: A review of local approaches in London'. In OECD (ed.), *From immigration to integration: Local solutions to a global challenge*, pp. 189–238. Paris: OECD.

Green, A. E., Owen, D., & Jones, P. (2007). *The economic impact of migrant workers in the West Midlands*. Birmingham, UK: West Midlands Regional Observatory.

Green, A. E., & Turok, I. (2000). 'Employability, adaptability and flexibility: Changing labour market prospects'. *Regional Studies*, 34(7): 599–600.

Greenham, F., with Moran, R. (2006). 'Complexity and community empowerment in regeneration'. In B. Temple & R. Moran (eds.), *Doing research with refugees: Issues and guidelines*, pp. 111–32. Bristol, UK: Policy Press.

Guiraudon, V. (2003). 'The constitution of a European immigration policy domain: A political sociology approach'. *Journal of European Public Policy*, 10(2): 263–82.

Habermas, J. (1981). *The theory of communicative action*, Thomas McCarthy, trans. (2 vols). Cambridge: Polity, 1984–87. (Trans. from *Theorie des kommunikativen Handelns*, 2 vols. Frankfurt am Main: Suhrkamp, 1981.)

Halde (2007). *Rapport annuel 2006*. Haute Autorité de Lutte Centre les Discriminations et pour l'Egalité Paris (HALDE).

Hall, P. A., & Soskice, D. (2001). 'An introduction to varieties of capitalism'. In P. A. Hall & D. Soskice (eds.), *Varieties of capitalism*, pp.1–68. Oxford: Oxford University Press.

Hammersley, G., & Johnson, J. (2004). *The experiences and perceptions of applicants who pursue claims at employment tribunals*. Paper presented at Work, Employment & Society Conference, UMIST, 1–3 September 2004.

Haney-Lopez, I. (1996). *White by law: The legal construction of race*. New York: New York University Press.

Haque, R. (2003). *Migrants in the UK: A descriptive analysis of their characteristics and labour market performance, based on the Labour Force Survey*. Department for Work and Pensions, London.

Hardill, I. (2004). 'Transnational living and moving experiences: Intensified mobility and dual-career households'. *Population, Space and Place*, 10: 375–89.

Harding, J. (2000). *The uninvited: Refugees at the rich man's gate*. London: Profile Books.

Harris, H. (2004). *The Somali community in the UK: What we know and how we know it*. London: ICAR, www.icar.org.uk/content.proj.prs.html.

Harris, J., & Roberts, K. (2006). 'Challenging barriers to participation in qualitative research: Involving disabled refugees'. In B. Temple & R. Moran (eds.), *Doing research with refugees: Issues and guidelines*, pp. 155–66. Bristol, UK: Policy Press.

Hasluck, C., & Green, A. E. (2007). *What works for whom?: A review of evidence and meta-analysis*. Department for Work and Pensions Research Report 407, London.

Hawthorne, L. (2006). *Labour market outcomes for migrant professionals: Canada and Australia compared—executive summary.* Report prepared for Citizenship and Immigration Canada, Ottawa.

Hiebert, D. (1999). 'Local geographies of labour market segmentation: Montreal, Toronto, and Vancouver'. *Economic Geography,* 75, 4, 339–69.

Hiebert, D. (2000). 'Immigration and the changing Canadian city'. *Canadian Geographer,* 44, 1, 25–43.

Hiebert, D. (2002). 'The spatial limits to entrepreneurship: Migrant entrepreneurs in Canada'. *Tijdschrift-voor-Economische-en-Sociale-Geografie,* 93, 2,173–90.

Hiebert, D. (2003). *Are migrants welcome? Introducing the Vancouver community studies survey.* Vancouver Centre of Excellence Research on Immigration and Integration in the Metropolis, Working Paper Series No. 03-06.

Hjarno, J. (2003). *Illegal immigrants and development in employment in the labour markets of the EU.* Aldershot, UK: Ashgate.

Ho, C., & Alcorso, C. (2004). 'Migrants and employment: Challenging the success story'. *Journal of Sociology,* 40(3): 237–59.

Ho, S. Y., & Henderson, J. (1999). 'Locality and the variability of ethnic employment in Britain'. *Journal of Ethnic and Migration Studies,* 25: 2, 323–33.

Home Office. (2000). *Full and equal citizens: A strategy for the integration of refugees into the United Kingdom.* London: HMSO.

Home Office. (2002). *Secure borders, safe haven–integration with diversity in modern Britain,* white paper CM5387. London: Home Office.

Home Office. (2005). *Controlling our borders: Making migration work for Britain, five year strategy for asylum and immigration.* London: Home Office.

Home Office. (2005). *Integration matters: A national strategy for refugee integration.* London: Home Office.

Home Office. (2005b). *Sizing the unauthorised (illegal) migrant population in the United Kingdom in 2001.* Home Office Online Report 29/05. London: Home Office.

Home Office. (2006). *A points-based system: Making immigration work for Britain,* Cm 6741. Norwich, UK: The Stationery Office.

Home Office. (2006b). *Borders, immigration and identity action plan.* London: Home Office.

Home Office. (2007). *Enforcing the rules: A strategy to ensure and enforce compliance with our immigration laws.* London: Home Office.

Home Office & DWP. (2007). *The economic and fiscal impact of immigration,* Cm 7237. Norwich, UK: The Stationery Office, http://www.unauk.org/election/electionhr.html.

Hudson, M., Phillips, J., Ray, K., et al. (2007). *Social cohesion in diverse communities.* York, UK: Joseph Rowntree Foundation.

Hudson, R. (2004). 'Conceptualizing economies and their geographies: Spaces, flows and circuits'. *Progress in Human Geography,* 28(4): 447–71.

Human Rights and Equal Opportunities Commission. (2004). *Isma—listen: National consultation on eliminating prejudice against Arab and Muslim Australians.* Croydon Park, NSW, Australia: HREOC.

Hurstfield, J., Pearson, R., Hooker, H., et al. (2004). *Employing refugees: Some organisations' experiences.* Brighton, UK: Institute for Employment Studies.

Hynes. P. (2003). 'The issue of trust and mistrust in research with refugees. Choices, caveats and considerations for researchers: New issues in refugee research', *Working Paper 98.* Geneva: UNHCR.

IDeA. (2007). *New European migration: Good practice guide for local authorities.* London: IDeA.

IES. (2004). *Employing refugees—some organisations' experience.* Brighton, UK: Institute of Employment Studies.

Institute of Community Cohesion. (2007). *Estimating the scale and impacts of migration at the local level.* London: Report for the Local Government Association.

Iredale, R., Mitchell, C., Pe-Pua, R., et al. (1996). *Ambivalent welcome: The settlement experiences of humanitarian entrant families in Australia.* Canberra: Bureau of Immigration, Multicultural and Population Research.

JCWI. (2006). *Recognising rights, recognising political realities: The case for regularizing irregular migrants.* London: Joint Council for the Welfare of Immigrants.

Jefferys, S. (2003). *Liberté, egalité and fraternité at work: Changing French employment relations and management.* Basingstoke, UK: Palgrave Macmillan.

Jefferys, S. (2004). *Racism and trade unions in the EU: National perspectives on a common challenge.* London: Working Lives Research Institute.

Jefferys, S. (2006). *Ambiguous messages: The gap between European trade union policies and the challenge of racism and xenophobia at the workplace.* London: Working Lives Research Institute.

Jenkins, R. (2004). *Social identity* (2nd ed.), London: Routledge.

Joint Standing Committee on Migration. (2004). *To make a contribution: Review of skilled labour migration programmes.* Canberra: Parliament of the Commonwealth of Australia .

Joly, D. (1996). *Haven or hell: Asylum policy in Europe.* London: Macmillan.

Jordan, B., & Düvell, F. (2002). *Irregular migration: The dilemmas of transnational mobility.* Cheltenham, UK: Edward Elgar.

Jordan, B., Stråth, B., & Triandafyllidou, A. (2003a). 'Comparing cultures of discretion'. *Journal of Ethnic and Migration Studies,* 29(2): 373–95.

Jordan, B., Stråth, B., & Triandafyllidou, A. (2003b). 'Contextualising immigration policy implementation in Europe'. *Journal of Ethnic and Migration Studies,* 29(2): 195–224.

Jupp, J. (2002). *From white Australia to Woomera: The story of Australian immigration.* Cambridge: Cambridge University Press.

Kelly, P. F., & D'Addario, S. (2008). 'Filipinos are very strongly into medical stuff: Labour market segmentation in Toronto, Canada'. In J. Connell (ed.), *The international migration of health workers,* pp. 77–96. New York: Routledge.

Kempton, J. (2002). *Migrants in the UK: Their characteristics and labour market outcomes and impacts,* RDS Occasional Paper 82. London: Home Office.

King, R. (1986). *Return migration and regional economic problems.* London: Croom Helm.

King, R. (2002). 'Towards a new map of European migration'. *International Journal of Population Geography,* 8(2): 89–106.

King, R., & Ruiz-Gelices, E. (2003). 'International student migration and the European "Year Abroad": Effect on European identity and subsequent migration behaviour'. *International Journal of Population Geography,* 9: 229–52.

Kirk, R. (2004). *Skills audit of refugees.* Home Office Online Report 37/04, http://www.homeoffice.gov.uk/rds/pdfs04/rdsolr3704/pdf.

Kissoon, P. (2006). 'Homelessness as an indicator of integration: Interviewing refugees about the meaning of home and accommodation'. In B. Temple & R. Moran (eds.), *Doing research with refugees: Issues and guidelines,* pp. 75–96. Bristol, UK: Policy Press.

Knapik, M. (2006). 'The qualitative research interview: Participants responsive participation in knowledge making. *International Journal of Qualitative Methods,* 5(3) 1–13.

Knowles, C. (2006). 'Handling your baggage in the field of reflections on research relationships'. *International Journal of Social Research Methodology,* 9(5): 393–404.

Knox, K. (1997). *A credit to the nation: A study of refugees in the United Kingdom.* London: Refugee Council.

Koch, W. (2006, April 25). 'Mixed status tears apart families'. *USA Today*. Retrieved January 5, 2008, at http://www.usatoday.com/news/nation/2006-04-25-mixed -status_x.htm.

Kochhar, R. (2005). *The occupational status and mobility of Hispanics*. Washington, DC: Pew Hispanic Center.

Kochhar, R. (2007b). '1995–2005: Foreign-born Latinos make progress on wages'. Washington, DC: Pew Hispanic Center.

Kofman, D., Raghuram, P., & Merefield, M. (2005). 'Gendered migrations—towards gender sensitive policies in the UK'. *Asylum and Migration Working Paper 6*. London: Institute for Public Policy Research

Kofman, E. (2004). 'Gendered global migrations'. *International Feminist Journal of Politics*, 6(4): 643–65.

Kofman, E. (2004). 'Family-related migration: Critical review of European studies'. *Journal of Ethnic and Migration Studies*, 30(2): 243–63.

Kofman, E., & Raghuram, P. (2006). 'Gender and global labour migrations: Incorporating skilled workers'. *Antipode*, 38, 2, 282–303.

Kohli, M. (1986). 'Social organisation and subjective construction of the life course'. in A. B. Sorenson, F. E. Weiner & L. R. Sherrod (eds.), *Human development and the life course*, pp. 271–92. Hillsdale, NJ: Lawrence Erlbaum Associates.

KPMG. (2005). *KPMG Review of English for speakers of other languages (ESOL)*. Report for the Department for Education and Skills for Life Strategy Unit and the Learning and Skills Council, Coventry.

Kuepper, W. G., et al. (1975). *Ugandan Asians in Great Britain: Forced Migration and Social Absorption*. London: Croom Helm.

Kunz, J. L., Milan, A., & Schetagne, S. (2000). *Unequal access. A Canadian profile of racial differences in education, employment and income*. A Report prepared for Canadian Race Relations Foundation by the Canadian Council on Social Development, Ottawa.

Lamba, N. K. (2003). 'The employment experience of Canadian refugees: Measuring the impact of human and social capital on quality of employment'. *The Canadian Review of Sociology and Anthropology*, 40(1): 45–64.

Lammers, E. (2005/April). 'Refugees, asylum seekers and anthropologists: The taboo on giving'. *Global Migration Perspectives, Paper No. 29*, Global Commission on International Migration.

Larsen, J. A., Allan, H. T., Bryan, K., et al. (2005). 'Overseas nurses' motivations for working in the UK: Globalization and life politics'. *Work, Employment and Society*, 19(2): 349–68.

Lave, J., & Wenger, E. (1991). *Situated learning: Legitimate peripheral participation*. Cambridge: Cambridge University Press.

Levinson, A. (2005). *The regularisation of unauthorized migrants: Literature survey and country case studies*. Oxford: Centre on Migration, Policy and Society, University of Oxford.

Li, P. S. (1999). 'Economic returns of migrants' self-employment'. *Canadian Journal of Sociology*, 25, 1, 1–34.

Lloyd, C. (2000). 'Trade unions and immigrants in France: From assimilation to anti-racist networking'. In R. Pennix & J. Roosblad (eds.), *Trade unions, immigration, and immigrants in Europe, 1960–1993*. Oxford: Berghahn Books.

Lopes, S. (2004). *Bringing employers into the immigration debate: The public policy implications of the survey findings*. A report prepared by the Public Policy Forum, Ottawa.

Los Angeles Times/Bloomberg. (2006). Poll. June 24–27.

Lund, M. (2006). *Effective methods in introductory programmes for immigrants*. Oslo: FAFO.

Marchand, O., & Thélot, C. (1991). *Deux siècles de Travail en France*. Paris: INSEE.

Martens, A. (1999). 'Migratory movements: The position, the outlook: Charting a theory and practice for trade unions'. In J. Wrench & N. Ouali (eds.), *Migrants, ethnic minorities and the labour market*. Basingstoke, UK: Macmillan.

Marx, K. (1964). *Il Capitale. Libro I, sezione 7, cap. 23*. Roma: Editori Riuniti.

Massey, D. (1987). 'Do undocumented immigrants earn lower wages than legal immigrants?: New evidence from Mexico'. *International Migration Review*, 21: 236–74.

Massey, D. S., Arango, J., & Husk, G. (1998). *Worlds in motion—understanding international migration at the end of the millennium*. OUP Oxford.

Matthews, G., & Ruhs, M. (2007). *Are you being served? Employer demand for migrant labour in the UK's hospitality sector*, COMPAS Working Paper 07-51. Oxford: University of Oxford.

mbA. (1999). *Creating the conditions for refugees to find work*. London: Report for the Refugee Council.

McCabe, A., Goodwin, P., & Garry, K. (2006). *JobCentre Plus and refugees: Progress in implementing the refugee employment strategy*. Employability Forum.

McCall, M. (1997). *High fliers*. Boston: Harvard Business School Press.

McKay, S., Chopra, D., & Craw, M. (2006a). *Migrant workers in England and Wales: An assessment of migrant worker health and safety risks*. Health and Safety Executive, London.

McKay, S., Dhudwar, A., & El Zailaee, S. (2006). *Comparing the labour market experiences of refugees and ethnic minorities*. London: Working Lives Institute, London Metropolitan University.

McKay, S., & Winkelmann-Gleed, A. (2005). *Migrant workers in the east of England*. Project report. EEDA, Norwich.

McLaughlan, G., & Salt, J. (2002). *Migration policies toward highly skilled foreign workers*. London: University College London, Migration Research Unit, Report to the Home Office.

Meldolesi, L. (1972). *Disoccupazione ed esercito industriale di riserva in Italia*. Bari, Italy: Laterza.

Menz, G. (2002). 'Patterns in EU labour immigration policy: National initiatives and European responses'. *Journal of Ethnic and Migration Studies*, 28(4): 723–42.

Mestheneos, E. (2006). 'Refugees as researchers: Experiences from the project "Bridges and fences: Paths to refugee integration in the EU"'. In B. Temple & R. Moran (eds.), *Doing research with refugees: Issues and guidelines*, pp. 21–36. Bristol, UK: Policy Press.

Mestheneos, E., & Ioannidi, E. (2002). 'Obstacles to refugee integration in the European Union member states'. *Journal of Refugee Studies* (15)3: 304–20.

Meurs, D., Pailhé, A., et al. (2005). 'Mobilité intergénérationnelle et persistence des inégalités'. *Documents de Travail INED* Vol. 130, No. 26.

Meyers, D. W. (2006). 'Temporary worker programs: A patchwork policy response'. *Insight*. Washington, DC: Migration Policy Institute.

Mezzadra, S. (2001). *Diritto di fuga, migrazioni, cittadinanza, globalizzazione*. Verona, Italy: Ombre Corte.

Midlands Refugee Council. (2001). *Asylum seekers: Developing information, advice and guidance on employment, training and education in the West Midlands*. EQUAL.

Migrationwatch. (2005). 'The illegal migrant population in the UK'. *Briefing Paper* 9.15, http://www.migrationwatchuk.org/Briefingpapers/migration_trends/illegal_migrant_pop_in_uk.asp.

Modood, T. (2004). 'Capitals, ethnic identity and educational qualifications. *Cultural Trends*, 13(2), No. 50: 87–105.

Mondi, F. (2005). *Pathways to employment*. London: RETAS.

Moran, R. A., & Butler, D. S. (2001). 'Whose health profile?' *Critical Public Health*, 11(1): 59–74.

Morrice, L. (2007). 'Lifelong learning and the social integration of refugees in the UK: The significance of social capital'. *International Journal of Life Long Education*, 26(2): 155–72.

Morris, L. (2001). 'The ambiguous terrain of rights: Civic stratification in Italy's emergent immigration regime'. *International Journal of Urban and Regional Research*, 25(3): 497–516.

Morris, L. (2002). 'Britain's asylum and immigration regime: The shifting contours of rights'. *Journal of Ethnic and Migration Studies*, 28(3): 409–25.

Morris, L. (2003). 'Managing contradiction: Civic stratification and migrants' rights'. *International Migration Review*, 37(1): 74–100.

Mottura, G. (2002). *Non solo braccia: Condizioni di lavoro e percorsi di inserimento sociale degli immigrati in un area ad economia diffusa*. Modena, Italy: Università di Modena e Reggio Emilia, Dipartimento di Economia Politica.

Mottura, G. (2003). 'Necessari ma non garantiti. I fattori di vulnerabilità socio-economica presenti nella condizione di immigrato'. In F. Carchedi, G. Mottura & E. Pugliese (eds.), *Il lavoro servile e le nuove schiavitù*, pp. 61–82. Milano: FrancoAngeli.

Mottura, G. (2006). 'Le badanti come nuove figure sociali'. *Quaderni Rassegna Sindacale*, 3: 91–104.

Mottura, G. (ed.) (1992). *L'arcipelago immigrazione: Caratteristiche e modelli migratori dei lavoratori stranieri in Italia*. Roma: Ediesse.

Mottura, G., & Pugliese, E. (1975). *Agricoltura, Mezzogiorno e mercato del lavoro*. Bologna: Il Mulino.

Mottura, G., & Pugliese, E. (2005). 'Presenza straniera e società in Italia: Il caso delle badanti'. In L. Dicomite, V. Rodriguez & S. Girone, S. (eds.), *Sviluppo demografico e mobilità territoriale delle popolazioni nell'area del Mediterraneo: Italia e Spagna a confronto*, pp.179–205. Bari, Italy: Cacucci.

Mottura, G., & Rinaldini, M. (2004). 'La precarietà autorizzata e non. Lavoratori stranieri nel mercato italiano'. *La Rivista delle Politiche Sociali*, 3: 217–36.

Munhall, P. (1988). 'Ethical considerations in qualitative research'. *Western Journal of Nursing*, 10: 150–62.

Myers, M. D. (1997). 'Qualitative research in information systems'. *MIS Quarterly*, 21:2: 241–42. MISQ Discover, archival version, June 1997.

National Refugee Integration Forum Employment and Training Subgroup. (2006). *Rebuilding lives—groundwork: Progress report on refugee employment*. London: Home Office and National Refugee Integration Forum.

NIACE. (2006). *'More than a language . . .' Interim report of the NIACE Committee of Inquiry on English for speakers of other languages*'. Leicester, UK: National Institute of Adult Continuing Education.

Noiriel, G. (1986). Les ouvriers dans la societe francaise: xix–xx siecle. Paris: Editions du Seuil.

NOP Business, Institute for Employment Studies. (IES) (2002). *Knowledge migrants*. London: Department of Trade and Industry (DTI) and the Home Office.

North, D. S., & Houston, M. (1976). *The characteristics and role of illegal aliens in the United States labor market: An exploratory study*. Washington, DC: Mimeo, Linton & Co.

Nun, J. (1970). 'Sobrepoblation relativa, ejército industrial de reserva y masa marginal'. *Revista Latino Americana de Sociologia*, 5(2) 47–66.

Oakley, A. (1981). 'Interviewing women: A contradiction in terms'. In Helen Roberts (ed.), *Doing feminist research*, pp. 83–113. London: Routledge and Kegan Paul.

Ong, A. (1999). *Flexible citizenship: The cultural logics of transnationality*. Durham, NC: Duke University Press.

ONS. (2004). *International migration—migrants entering or leaving the UK and England and Wales, 2003.* London: Office for National Statistics.

Opengart, R., & Short, D. C. (2002). 'Free agent learners: The new career model and its impact on human resource development'. *International Journal of Lifelong Education,* 21(1): 220–33.

Oppenheim, A. (1992). *Questionnaire design, interviewing and attitude measurement.* London: Pinter.

Ossman, S. (2004). 'Studies in serial migration'. *International Migration,* 42(4): 111–21.

Pastor, M., & Alva, S. (2004). 'Guest workers and the new transnationalism: Possibilities and realities in an age of repression'. *Social Justice,* 31: 1–2.

Patton, M. Q. (2002). *Qualitative research and evaluation methods.* Thousand Oaks, CA: Sage.

Pendakur, K., & Pendakur, R. (2002). 'Speaking in tongues: Language, knowledge as human capital and ethnicity. *International Migration Review,* Spring, 36, 1,147–78.

Pendakur, R. (2000). *Migrants and the labour force: Policy, regulation, and impact.* Montreal: McGill-Queen's University Press.

Pew Hispanic Center. (2006a/April). 'Estimates of the unauthorized migrant population for states based on the March 2005 CPS'. Washington, DC.

Pew Hispanic Center. (2006b/October). 'Populaton by race and ethnicity: 2005'. Washington, DC.

Pew Hispanic Center. (2006c/October). 'Occupation by region of birth: 2005'. Washington, DC.

Pew Hispanic Center. (2006d/October). 'Personal earnings by region of birth: 2005'. Washington, DC.

Pew Hispanic Center. (2006e/October). 'Poverty by age and region of birth: 2005'. Washington, DC.

Pew Hispanic Center. (2007/December). 'The immigration debate: Controversy heats up, Hispanics feel chill'. Washington, DC.

Phillimore, J. (2005). *Evaluating the outcomes of refugee support.* Birmingham, UK: Report for Birmingham City Council.

Phillimore, J., & Goodson, L. (2001). *Exploring the integration of asylum seekers and refugees in Wolverhampton into UK labour market.* Birmingham, UK: Centre for Urban and Regional Studies, University of Birmingham.

Phillimore, J., & Goodson, L. (2002). *Asylum seeker and refugee employability initiatives: Models for implementing a super pathway in the West Midlands.* Discussion paper prepared for West Midlands Executive Consortia.

Phillimore, J., & Goodson, L. (2006). 'Problem or opportunity? Asylum seekers, refugees, employment and social exclusion in deprived urban areas'. *Urban Studies,* 43(10): 1715–36.

Phillimore, J., Goodson, L., & Beebeejaun, Y. (2004). *The access, learning and employment needs of newcomers from abroad and the capacity of existing provision to meet those needs.* LSC Birmingham and Solihull.

Phillimore, J., Goodson, L., Oosthuizen, R., et al. (2003). *Asylum seekers and refugees: Education, training, employment, skills and services in Coventry and Warwickshire.* LSC Coventry and Warwickshire.

Phillimore, J. & Goodson, L. (2008). *New Migrants in the UK: Education, Training, and Employment.* London: Trentham.

Picot, G., Hou, F., & Coulombe, S. (2007). *Chronic low income and low-income dynamics among recent migrants.* Analytical Studies Branch Research Paper Series No. 294. Ottawa: Statistics Canada.

PICUM. (2006). *Comments on the communication from the commission on 'policy priorities in the fight against illegal immigration of third-country nationals'.* Brussels: Platform for International Co-operation on Undocumented Migrants.

Piore, M. (1979). *Birds of passage: Migrant labor in industrial societies*. Cambridge: Cambridge University Press.

Portes, A., & Bach, R. L. (1985). *Latin journey: Cuban and Mexican immigrants in the United States*. Berkeley and Los Angeles: University of California Press.

Portes, J., & French, S. (2005). *The impact of free movement of workers from Central and Eastern Europe on the UK labour market: Early evidence*. Department for Work and Pensions, Working Paper 18 London.

Pratt, G. (1999). 'From registered nurse to registered nanny: Discursive geographies of Filipina domestic workers in Vancouver, B.C.'. *Economic Geography*, 75, 3, 215–36.

Preston, V. (2004). *Employment experiences of highly skilled migrant women: Where are they in the labour market?* A paper presented at the "Gender & Work: Knowledge Production in Practice" Conference, October 1–2. Toronto: York University.

Preston, V., & Cox, J. C. (1999). 'Migrants and employment: A comparison of Montreal and Toronto between 1981 and 1996'. *Canadian Journal of Regional Science*, 22, 1–2, 87–111.

Preston, V., & Giles, W. (1996). 'Ethnicity, gender and labour markets in Canada: A case study of migrant women in Toronto'. *Canadian Journal of Urban Research*, 6, 2, 135–47.

Preston, V., & Man, G. (1999). 'Employment experiences of Chinese migrant women: An exploration of diversity'. *Canadian Women's Studies*, 19, 115–22.

Pugliese, E. (1996). 'L'immigrazione'. In *Storia dell'Italia repubblicana. L'Italia nella crisi mondiale: L'ultimo ventennio*, 3(1): 933–84. Torino: Einaudi.

Pugliese, E. (2002). *L'Italia tra immigrazioni internazionali e migrazioni interne*. Bologna: Il Mulino.

Raghuram, P., & Kofman, E. (2002). 'The state, skilled labour markets, and immigration: The case of doctors in England'. *Environment and Planning A*, 34: 2071–89.

Refugee Council. (2005). *The government's five year asylum and immigration strategy*. London: Refugee Council Briefing.

Refugee Resettlement Working Group. (1994). *Let's get it right in Australia*. Sydney: Refugee Resettlement Working Group.

Regini, M. (2000). 'The dilemmas of labour market regulation'. In G. Esping-Anderson and M. Regini (eds.). *Why Deregulate Markets*, pp. 11–29, Oxford: OUP.

Reitz, J. G. (1998). *Warmth of the welcome: The social causes of economic success for migrants in different nations and cities*. Boulder, CO: Westview Press.

Reitz, J. G. (2002). 'Host societies and the reception of migrants: Research themes, emerging theories and methodological issues'. *International Migration Review*, 36, 1005–19.

Reyneri, E. (1998). 'The role of the underground economy in irregular migration to Italy: Cause or effect?' *Journal of Ethnic and Migration Studies*, 24: 313.

Reyneri, E. (2004). 'Education and the occupational pathways of migrants in Italy'. *Journal of Ethnic and Migration Studies*, 30(6): 1145–62.

Richardson, S., et al. (2004). *The changing labour force experience of new migrants*. The National Institute of Labour Studies, Flinders University, Adelaide (Australia). Report to the Department of Immigration and Multicultural and Indigenous Affairs, June 2004, at http://www.immi.gov.au/media/publications/pdf/labour-forcev2.pdf.

Richmond, A. (1993). 'Reactive migration: Sociological perspectives on refugee movements'. *Journal of Refugee Studies* (6)7–24.

Rinaldini, M. (2004). 'Immigrati e cittadini'. In I. Masulli (ed.), *Precarietà del lavoro e società precaria nell'Europa contemporanea*, pp. 145–58. Roma: Carocci.

Rivera-Batiz, F. L. (1999). 'Undocumented workers in the labor market: An analysis of the earnings of legal and illegal Mexican immigrants in the United States'. *Journal of Population Economics*, 12: 1: 91–116.

Robinson, V. (1998). 'Cultures of ignorance, disbelief and denial: Refugees in Wales'. *Journal of Refugee Studies*, 12(1): 78–87.

Robinson, V. (1999). 'The importance of information in the resettlement of refugees in the UK'. *Journal of Refugee Studies*, 11(2): 146–60.

Robinson, V., & Segrott, J. (2002/July). *Understanding the decision-making of asylum seekers*. Migration Unit, Department of Geography, University of Wales, Swansea. Home Office Research Study 243.

Rosenthal, F. (1993). 'Reconstruction of life stories: Principles of selection in generating stories for narrative biographical interviews'. In R. Josselson & A. Lieblich (eds.), *The narrative study of lives*. pp. 59–91. Newbury Park, CA: Sage.

Ruhs, M. (2005). *The potential of temporary migration programmes in future international migration policy*. Oxford: A paper prepared for the Policy Analysis and Research Programme of the Global Commission on International Migration.

Ruhs, M., & Anderson, B. (2006). *Semi-compliance in the migrant labour market*. COMPAS Working Paper, 1 May 2006, www.compas.ox.ac.uk/changing status.

Rydgren, J. (2004). 'Mechanisms of exclusion: Ethnic discrimination in the Swedish labour market'. *Journal of Ethnic and Migration Studies*, 30(4): 697–717.

Sales, R. (2002). 'The deserving and the undeserving? Refugees, asylum seekers and welfare in Britain'. *Critical Social Policy*, 22(3): 456–78.

Salt, J. (2005). *Types of migration in Europe: Implications and policy concerns*. European Population Conference.

Salt, J. (2006). *International migration and the United Kingdom*. Report of the United Kingdom SOPEMI Correspondent to the OECD.

Salt, J., Clarke, J., Dobson, J., et al. (2003). *The ins and outs of migration*. Commission for Racial Equality London.

Salt, J., & Millar, J. (2006/October). 'Foreign labour in the United Kingdom: Current patterns and trends'. *Labour Market Trends*, 335–55.

Sargeant, G., & Forna, A. (2001). *A poor reception: Refugees and asylum seekers: Welfare or work?* Policy Paper. London: The Industrial Society.

Saxenian, A. L. (2006). *The new Argonauts*. Cambridge, MA: Harvard University Press.

Schellenberg, G., & Maheux, H. (2007). 'Migrants' perspectives on their first four years in Canada: Highlights from three waves of the Longitudinal Survey of Migrants to Canada'. *Canadian Social Trends*, Special Edition, 2–17.

Schor, R. (1996). *Histoire de l'immigration en France de la fin du XIXe siècle à nos jours*. Paris: Armand Colin.

Scottish Executive. (2005). *Refugees and asylum seekers in Scotland: A skills and aspirations audit* Social Justice Research Programme Research Findings No 10/2004, Edinburgh: Scottish Executive.

Scottish Refugee Council. (2001). *Responding to the needs of asylum seekers*. Glasgow: Scottish Refugee Council.

Sennett, R. (2000). *The corrosion of character: The personal consequences of work in the new capitalism*. London: Norton.

Shields, M. A., & Wheatley Price, S. (2003). *The labour market outcomes and psychological well-being of ethnic minority migrants in Britain*. London: Home Office Online Report 07/03.

Shih, J. (2002). '"... Yeah, I could hire this one, but I know it's gonna be a problem": How race, nativity and gender affect employers' perceptions of the manageability of job seekers'. *Ethnic and Racial Studies*, 25(1): 99–119.

Social Trends. (2002). *Unemployment rates: By region, 2002: Social Trends 33*, 21/01/03

Solé, C., & Parella, S. (2003). 'The labour market and racial discrimination in Spain'. *Journal of Ethnic and Migration Studies*, 29(1): 121–40.

Somerville, W. (2007). *Immigration under New Labour*. Bristol, UK: Policy Press.

Spalek, B. (2005). 'A critical reflection on researching black Muslim women's lives post September 11th'. *International Journal of Social Research Methodology*, 8(5): 405–18.

Sriskandarajah, D., Cooley, L., & Reed, H. (2005). *Paying their way: The fiscal contribution of immigrants in the UK*. Institute for Public Policy Research London.

Statistics Canada. (2003). 'Update on cultural diversity'. *Canadian Social Trends*, Autumn, 19–23.

Stevens, D. (2003). 'Roma Asylum Applicants in the United Kingdom: "Scroungers" or "Scapegoats"'., In Joanne van Selm et al. (eds.), *The refugee convention at fifty: A view from forced migrations studies*, pp. 145–60. MD: Lexington Books New York and London.

Stewart, E. (2003). 'A bitter pill to swallow: Obstacles facing refugee and overseas doctors in the UK'. *New issues in refugee research*, Working Paper 96. UNHCR Evaluation and Policy Analysis Unit Geneva:

Sum, A., Harrington, P., & Khatiwada, I. (2006/September). 'The impact of new immigrants on young native-born workers, 2000–2005'. Washington, DC: Center for Immigration Studies.

Tait, K. (2006). 'Refugee voices as evidence in policy and practice'. In B. Temple & R. Moran (eds.), *Doing research with refugees: Issues and guidelines*, pp 133–53. Bristol, UK: Policy Press.

Tastsoglou, E., & Preston, V. (2005). 'Gender, immigration and labour market integration: Where we are and what we still need to know'. *Atlantis*, 30, 1, 46–59.

Temple, B., & Moran, R. (edss) (2006). *Doing research with refugees: Issues and guidelines*. Bristol, UK: Policy Press.

Temple, Bogusia, & Edwards, R. (2006b). 'Limited exchanges: Approaches to involving people who do not speak English in research and service development'. In B. Temple & R. Moran (eds.), *Doing research with refugees: Issues and guidelines*, pp 37–54. Bristol, UK: Policy Press.

Thalhammer, E., Zucha, V., et al. (2001). *Attitudes towards minority groups in the European Union*. Vienna: European Monitoring Centre on Racism and Xenophobia.

Thierry, X. (2004). 'Recent immigration trends in France and elements for a comparison with the United Kingdom'. *Population* 59(5): 635–72.

Thompson, S. (1996). 'Paying respondents and informants'. *Social Research Update 40*. Surrey, UK: University of Surrey, www.soc.surrey.ac.uk/sru.

Tilbury, F., & Colic-Peisker, V. (2006). 'Deflecting responsibility in employer talk about race discrimination'. *Discourse and Society*, 17: 651–76.

Triandafyllidou, A. (2003). 'Immigration policy implementation in Italy: Organisational culture, identity processes and labour market control'. *Journal of Ethnic and Migration Studies*, 29(2): 257–97.

Tripier, M. (1990). *L'immigration dans la classe ouvrière en France*. University of Nantes, Paris:

Tuitt, P. (1996). *False images: Law's construction of the refugee*. London: Pluto Press.

U.S. Department of Homeland Security. (2004). *Yearbook of immigration statistics*. Washington, DC: Homeland Security.

U.S. General Accounting Office. (2000). *H-2A agricultural guestworkers: Status of changes to improve program services*. GAO/T-HEHS-00-134. Washington, DC.

UNA-UK. (2005). *The UK and the UN: Human Rights.* http://www.unawc.org/election/electionhr.html. London: United Nations Association.

URMIS. (2003). *National briefing: France.* Nice: Unité de Recherche Migrations et Société.

US Today/Gallup Poll. (2006). Online at http://www.usatoday.com/news/nation/2006-04-10-immigration-divide_x.htm. Accessed April 12.

Valtonen, K. (1999). 'The societal participation of Vietnamese refugees: Case studies in Finland and Canada'. *Journal of Ethnic and Migration Studies*, 25(3): 469–91.

Valtonen, K. (2004). 'From the margin to the mainstream: Conceptualizing refugee settlement process'. *Journal of Refugee Studies*, 17: 70–96.

Van der Heijden, B. I. M. (2002). 'Individual career initiatives and their influence upon professional expertise development throughout the career'. *International Journal of Training and Development*, 6(2): 54–79.

Van Hear, N. (1998). *New diasporas: The mass exodus, dispersal and regrouping of migrant communities.* London: UCL Press.

Vasta, E. (2004). *Informal employment and immigrant networks: A review paper*, Centre on Migration, Policy and Society Working Paper No. 2. Oxford: University of Oxford.

Villiers, J. (1994). 'Closed borders, closed ports: The flight of Haitians seeking political asylum in the United States'. *Brooklyn Law Review*, 60: 841.

Vinci, S. (1975). *Il mercato del lavoro in Italia.* Milano: FrancoAngeli.

Vourc'h, F., & de Rudder, V. (2002). 'Discriminations ethnistes et racistes: Nommer, compter, corriger'. *Sida, Immigration et inégalités*, ANRS. Paris: Collection Sciences Sociales et Sida.

Walters, N., & Egan E. (1996). *Refugee skills analysis report for North West London Training & Enterprise Council.* University of Surrey, Guildford.

Waslander, B. (2003). 'Failing earnings of new migrants in Canada's cities'. In C. M. Beach, A. G. Green, & J. G. Reitz (eds.), *Canadian immigration policy for the 21st century*, pp. 335–372. Kingston, Ontario: John Deutsch Institute for the Study of Economic Policy, Queen's University.

Whitwell, C. (2002). *'New migration' in the 1990s: A retrospective*, Sussex Migration Working Paper No.13 Brighton: Sussex Centre for Migration Research.

Williams, A. (2001). 'New forms of international migration: in search of which Europe'. In H. Wallace (ed.), *Interlocking dimensions of European integration*, pp. 103–21. Basingstoke, UK: Palgrave.

Williams, A. M. (2006). 'Lost in translation? International migration, learning and knowledge'. *Progress in Human Geography*, 30(5): 588–607.

Williams, A. M., & Balaz, V. (2005). 'What human capital, which migrants? Returned skilled migration to Slovakia from the UK'. *International Migration Review*, 39(2): 439–68.

Williams, A. M., & Balaz, V. (2008). *International migration and knowledge.* London: Routledge.

Wood, G. A. (1990). 'Occupational segregation by migrant status in Australia'. *Economics Programme*, Working Paper No. 48. Perth, Australia: Murdoch University.

Wooden, M. (1991). 'The experience of refugees in the Australian labour market'. *International Migration Review*, 25(3): 514–35.

Wright, T. & Pollert, A. (2006). *The experience of ethnic minority workers in the hotel and catering industry: Routes to support and advice on workplace problems.* ACAS Research Paper, 03/06, London: ACAS.

Zetter, R., & Pearl, M. (2000). 'The minority within the minority: Refugee community-based organisations in the UK and the impact of restrictionism on asylum seekers'. *Journal of Ethnic and Migration Studies*, 26(4): 675–97.

Zetter, R., with Griffiths, D., Sigona, N., Flynn, S., et al. (2006). *Immigration, social cohesion and social capital: What are the links?* York, UK: Joseph Rowntree Foundation.

Zetter, R., & Pearl, M. (1999). *Managing to survive—asylum seekers, refugees and access to social housing.* Bristol, UK: Policy Press.

Zetter, R., with Pearl, M. (2005). *Still surviving and now settling: Refugees, asylum seekers and a renewed role for housing associations.* Oxford: Oxford Brookes University.

Zetter, R., Griffiths, D., Sigona, N., et al. (2002). *Survey on policy and practice related to refugee integration.* Oxford, European Refugee Fund Community Actions 2001/2002; conducted by School of Planning, Oxford Brookes University. DIAC (Department of Immigration and Citizenship) (2007) *Fact Sheet 66: Integrated humanitarian settlement strategy*, at http://www.immi.gov.au/media/fact-sheets/66ihss.htm.

Zincone, G. (2006). 'The making of policies: Immigration and immigrants in Italy'. *Journal of Ethnic and Migration Studies*, 32(3): 347–75.

Zulauf, M. (1999). 'Cross-national qualitative research: Accommodation ideals and reality. *International Journal of Social Research Methodology*, 2(2): 159–69.

Contributors

Dr Alice Bloch is a Senior Lecturer in the Department of Sociology at City University, London. Her research has mainly explored the economic and social settlement of refugees in the UK. More recently she has been researching the costs and benefits of migration for sending and receiving countries, including remittances and other transnational activities. Dr Bloch is currently writing a book on *Migration and Citizenship* for Palgrave with Liza Schuster and coediting a book with John Solomos, also for Palgrave, *Race and Ethnicity in the 21st Century*. Dr Bloch is course director for the MA in Refugee Studies at City University.

Val Colic-Peisker is a Senior Research Fellow/Senior Lecturer in the School of Global Studies, Social Science and Planning of the Royal Melbourne Institute of Technology (RMIT University, Australia). She is a sociologist/political scientist with special interest in immigration and settlement, labour market integration of immigrants and refugees in Australia, global mobility of professionals, urban studies and socio-cultural aspects of Australian housing market and home ownership. Val's recent publications include *Migration, class and transnational identities: Croatians in Australia and America* (2008) Urbana and Champaign: Illinois University Press and refereed articles in *Journal of Ethnic and Migration Studies* (2005), *Journal of Refugee Studies* (2006), *Discourse and Society*, (2006) *International Migration* (2007), *Journal of Intercultural Studies* (2008) and *Race and Class* (2008).

Silvia D'Addario is a Doctoral Student in the Department of Geography at York University. Her research areas include immigration and settlement in Canada and the geographies of residence and work for immigrants in contemporary suburbs. Her doctoral research examines the links between place of residence and workplace for transnational newcomers in the outer suburbs of Toronto, Ontario. Ms. D'Addario received her master's degree in geography from the University of British Columbia. Her MA research examined the housing experiences of refugee claimants in Vancouver, British Columbia.

Dr Anne Green is a Principal Research Fellow at the Institute for Employment Research, University of Warwick, has a background in geography and works primarily on issues concerned with spatial dimensions of economic, social and demographic change and on aspects of local and regional labour markets. Dr Green is vice-chair of the Regional Studies Association and a member of the Regional Science Association, the Royal Geographical Society (with the Institute of British Geographers) and the Town and Country Planning Association.

Professor Steve Jefferys is the Director of the Working Lives Research Institute, London Metropolitan University, and was previously Professor of European Employment Studies at London Metropolitan University. He has led a number of EU-funded research projects on migrant workers and was the coordinator of a five-country Framework 5 project on 'Racism in Trade Unions'.

Dr Sonia McKay joined the Working Lives Research Institute in 2004 to head an ESF-funded project on refugees and their labour market exclusion. She currently heads a number of research projects focusing on refugees and recent migrants, including a seven-partner Framework Six–funded project on Undocumented Workers Transitions. Sonia is an employment law specialist and had previously worked for the Labour Research Department (LRD), the independent trade union–based research organisation. Before that she had worked for a UK trade union for eight years.

Professor Giovanni Mottura holds a Chair in Sociology of Work at the University of Modena and Reggio Emilia. He is also the editor of the IRES-CGIL annual Observatory of Immigration report. Over more than two decades he has been researching on migration in Italy and Europe, mainly focusing on the labour market. He is the author of many articles and research reports in this area.

Dr Jenny Phillimore is a lecturer and course director Centre for Urban and Regional Studies, University of Warwick. Dr Phillimore has undertaken projects for sponsors including the LSC, DETR, the Home Office, The London Action Trust, The Welsh Assembly, Chambers of Commerce, the Housing Corporation and a number of different local authorities. These have examined issues assessing the needs and aspirations of asylum seekers and refugees, single parents, young Afro-Caribbean men, the disabled, long-term unemployed, drug and alcohol misusers, ex-offenders and young people at risk of offending in relation to education, training and employment. Her work has contributed to the development of policy recommendations covering excluded communities and specific groups including ethnic minorities, asylum seekers and refugees, ex-offenders, single parents, the disabled and young people. Together with Lisa Goodson she has recently published a book on qualitative research for Routledge.

Dr Valerie Preston is Professor of Geography at York University, where she teaches urban social geography. Her research focuses on the inclusion of newcomers in contemporary cities. Currently, she is exploring the settlement experiences of migrants in a Canadian suburb where high housing costs, limited public transport and low densities impede access to suitable and remunerative employment. With several collaborators, she is also completing a study of transnational migration and the citizenship practices and identities of Hong Kong immigrants in Canada. Focusing on the links between home and work, she has published numerous articles about women's employment in North American cities

Matteo Rinaldini is a PhD student working under Professor Mottura at the University of Modena and Reggio Emilia. In 2006 he was a visiting research scholar at the Working Lives Research Institute, London Metropolitan University. In Italy he worked on a number of research projects focusing on migration and is currently researching on migrant status and working conditions.

Dr Nandita Sharma is an Assistant Professor cross appointed in the Ethnic Studies Department and the Sociology Department, University of Hawai'i at Manoa. Her research interests are in the areas of international migration, national and international policies regarding migration, historic and contemporary processes of globalization, national state power, ideologies of racism and nationalism, and transnational social movements for justice. Dr Sharma is an activist scholar whose research is shaped by the social movements she is active in, including No Borders movements, feminist, antiracist, anticolonial movements, sex-workers' movements, and movements for ecologically sound practices

Paula Snyder is the producer and director of *All by Myself—Looking for Work as a Refugee*. Paula Snyder is a freelance filmmaker and journalist. She worked at Channel 4 for seven years commissioning all educational support for the main schedule, including award-winning campaigns like the *Black and Asian History Map* and *Brookie Basics*, and all telephone helplines.

Professor Allan Williams was appointed to the Chair in European Integration and Globalization at London Metropolitan University in 2006. He is a member of both the Institute for the Study of European Transformations and the Working Lives Research Institute. Prior to his appointment to London Metropolitan University he held a chair in Human Geography and European Studies at the University of Exeter. He is an Academician of the Academy of Social Science and has been a member of several Economic and Social Research Council (ESRC) committees, including the Research Grants Board 2001-5 and the 'One Europe or Several' Commissioning Panel, 1997–9. He chaired the ESRC/NERC Transdisciplinary Seminars competition in 2005, and is currently a member of the ESRC

First Grants Commissioning Panel. He is an adjunct professor in the National Centre for Research on Europe at the University of Canterbury in New Zealand, where he is developing a research programme on European–New Zealand migration. He is coeditor of two journals: *European Urban and Regional Studies* and *Tourism Geographies*. He is a member of the editorial boards of two other journals: *Mobilities* and *Annals of Tourism Research*.

Tessa Wright is a Senior Research Fellow at the Working Lives Research Institute and is currently completing her PhD at Queen Mary College. Tessa previously worked for the Labour Research Department as the department's equality researcher. Tessa is currently working on a Framework Six–funded project on Undocumented Workers Transitions. Prior to this Tessa had worked on two projects on the experience of ethnic minority workers in the hotel and catering industry and on multiple discrimination and social exclusion at work, exploring the intersection between gender, race, age and sexual orientation.

Index